fifth edition

SMALL FOOTPRINT BIG IMPACT

INTRODUCTION TO ENVIRONMENTAL SCIENCE

one-time online access code included

ELIZABETH A. JORDAN

Kendall Hunt
publishing company

Kendall Hunt
publishing company

www.kendallhunt.com
Send all inquiries to:
4050 Westmark Drive
Dubuque, IA 52004-1840

Textbook with Website ISBN 978-1-7924-1197-7
Textbook alone ISBN 978-1-7924-1198-4

Published in the United States of America

In loving memory of the June 7, 2013
Santa Monica College shooting victims.

Contributing Editors

A Thomas Jordan

Tylor Connor

Heather Hamza

Graeme Gissing

Contents

Chapter 1

Introduction to Environmental Science

Congratulations! You are reading this because you are determined to learn about the environment while making a positive difference in it. We would all like to help make our environment a cleaner, healthier place to live and conserve natural resources and wildlife habitats.

This seems like an overwhelming task, so where do we begin? How can we possibly preserve our natural environment and still maintain a quality lifestyle and gain benefits from our natural resources?

Although there are no simple answers to these questions, this textbook is designed to help you get started. We will discuss environmental science and offers suggestions to help you live more sustainably, lower your personal ecological footprint, and help conserve and clean up our environment.

This book touches upon such personally relevant and controversial topics as GMOs, pollution, conservation, global warming, renewable versus nonrenewable energy, and biodiversity. It tracks the progress and marked success of a galvanized international community committed to ensuring that no person or living being carries the unjust burden generated by overconsumption and unsustainable environmental practice.

My hope is that you will have a similar reaction. Seven billion people on the earth mean that there are 7 billion ways to make a positive impact. It starts with just one. So let us get you started on your environmentally “green” journey!

DID YOU KNOW?

Iceland: June 8, 1783. A volcano erupted and continued its activity over a span of 8 months. It released a gas cloud which killed half of the country's livestock and one-quarter of its native Icelanders. The deaths were a result of poisoning and famine (crops failed due to acid rain). This toxic nebulous cloud eventually made its way to France, Norway, and Germany, where it was linked to countless deaths. Countries as far as Syria experienced the adverse effects of the noxious air. In fact, the eruption triggered the most severe weather patterns in the northern hemisphere of the last 250 years which compelled the notable Benjamin Franklin himself to write about its harmful effects on weather.[i]

What was this gas that was responsible for such massive death and destruction? It is a gas called sulfur dioxide. The sulfur dioxide released from the explosion eventually mixed with water vapor to create acid rain that destroyed their crops. Unfortunately, this insidious gas also mixed with water vapor in the lungs of human victims, ultimately choking them.

What is the significance of sulfur dioxide? It is the gas produced by burning fossil fuel and released by coal-burning plants and other industrial processes. It causes respiratory problems and reacts with water vapor to produce acid rain. As the devastating episode in Iceland demonstrates, low intensity but long, protracted emission and high-volume pollution can have cataclysmic effects—rivaling that of a more dramatic, abrupt, and explosive natural disaster, even though human activity does not release sulfur dioxide on the scale of the referenced volcano. Is it possible for human activity and man-made pollution to mimic this tragic historic chapter in our recent history? Can our pollution create the same direct consequences as observed with natural disasters such as the volcanic eruption in Iceland? If so, what can we do to curb its adverse effects?

Morphart Creation/Shutterstock.com

What is Environmental Science?

Environmental science is a multidisciplined study of humans' interaction with the living and the nonliving world. Encompassing sociology, physics, ecology, geography, humanities, economics, climatology, chemistry, meteorology, political science, and many more, its goal is to examine and explore the many environmental issues that we face today.

The multiple and varied disciplines listed above suggest how intricate, involved, and multifaceted environmental science is and raises profound questions about their implications. For example:

- What accounts for the elevated wildlife extinction rate and what can we do to curb this rate?
- How do we provide a growing population with increasing energy demands and sustainable infrastructure?
- What are factors that go into successful conservation?
- How does an oil spill affect marine biodiversity of the Gulf of Mexico?
- What are the environmental concerns of a compromised nuclear reactor in Japan?
- How do socioeconomics affect a person's view on environmental issues and what they are willing to do about them?
- How does overgrazing of livestock impact desertification and greenhouse gas emissions?
- How do we make sure that everybody on the planet has access to clean water?
- How do we feed the earth's growing population without permanently stripping the soil of nutrients?
- How do we ensure that everybody has equal access and benefit to earth's biodiversity?

Cheryl Casey/Shutterstock.com

Each of these complex and thought-provoking scenarios fall under the umbrella of environmental science.

Terms to Know

As a species, we *Homo sapiens* have walked planet earth for less than 200,000 years but have, in this time, managed to negatively alter the earth and its natural resources to an unprecedented extent. Most of this change has occurred in the last 50 years.

In order to understand how to improve our environment, we must be able to understand the complexity of the living and nonliving worlds, which brings us to **ecology**. Not formally considered a science until as recently as the early 1900s, ecology is the study of the **biotic** (*living*) world and its interactions with the **abiotic** (*nonliving*) environment. The biotic world consists of plant life, wildlife, insects, pollinators, predators, fungi, as well as bacteria and earthworms whose job is to recycle dead matter and reintroduce it into the **ecosystem**. An ecosystem encompasses all living factors and how they interact with the physical and chemical (nonliving) environment. The abiotic environment is comprised of the following chemical and physical factors:

Chemical components of the environment include factors such as the gas molecules that make up earth's atmosphere as well as your city's local air quality. Chemical factors also address the composition of soil and water in a given ecosystem. Is the soil high or low in nutrient content, if so, what kind of nutrients? What is the water quality of an aquatic environment? Does it contain high or low salinity? Are there abundant concentrations of oxygen?

The **physical** environment addresses factors such as precipitation, temperature, and amount of sunlight. Does a given ecosystem receive abundant rainfall annually? What is its average yearly temperature? How many days out of the year experience average temperatures below the freezing point? These are all physical aspects of an ecosystem.

Now let's consider the **social** aspects of environmental science. **Environmentalism** is a social movement committed to conserving or improving the environment and living sustainably. To live **sustainably** is to live in such a way that vital goods are taken from the earth conservatively, avoiding long-term damage or depletion, while continuing to develop social and economic value in society. Sustainability is a concept whose goal is the best possible outcome for humans, wildlife, and the environment Sustainability also implies social justice. An **ecological footprint** is the amount of land and other natural resources required to supply a population with its goods and services while absorbing and incorporating its waste. The objective of environmentalism. is to help you reduce your personal ecological footprint. A **carbon footprint** is the amount of carbon (in the form of methane or carbon dioxide) released by a given **population**.

What Is Life?

It would take an entire philosophy class to even attempt to address the essence of "life." Instead of addressing "life" from a lofty philosophical lens, we will examine it from a scientific perspective. "Life" is the gestalt of the following conditions: life responds to external stimuli, has order, can reproduce, grow, evolve and adapt to the environment, breaks down its own food for energy, and maintains **homeostasis** (a consistent internal environment.) As you may notice, the aforementioned variables are indicators that lead us toward the *direction* of defining life.

Life—Small to Large

Now that we have the basics, let us track life's components from small to large: everything on this planet is made of atoms. An atom is the smallest unit of matter that retains specific properties. **Atoms** create bonds with each other to form the organic **molecules** of life: fats, carbohydrates, proteins, and DNA.

Organic molecules join together to create a cell, the basic unit of life. According to **The Cell Theory**, living things are made up of cells, which in turn are made up of component parts. In addition, all cells arise from preexisting cells that reproduce asexually, creating clones of themselves. Humans, mushrooms, plants, and bacteria are all made up of cells. Humans are

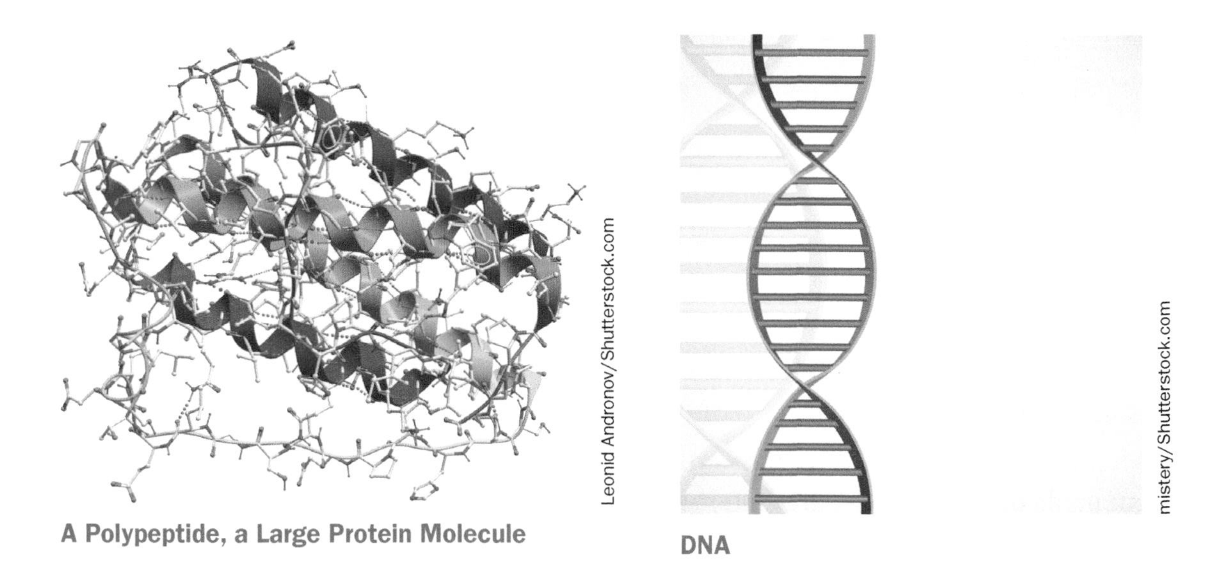
A Polypeptide, a Large Protein Molecule

Leonid Andronov/Shutterstock.com

DNA

mistery/Shutterstock.com

multicellular beings because we are made up of trillions of cells. Our cells each divide about 50–70 times then die, giving rise to newer cells. Some organisms such as an amoeba or bacterium are only one cell large! They are considered *unicellular*.

A group of different cell types that support each other for a common function is called tissue. Blood is made up of white blood cells that help fight disease-causing **pathogens**. Blood is also made up of red blood cells that carry oxygen to other parts of the body. Together, white and red blood cells form blood tissue.

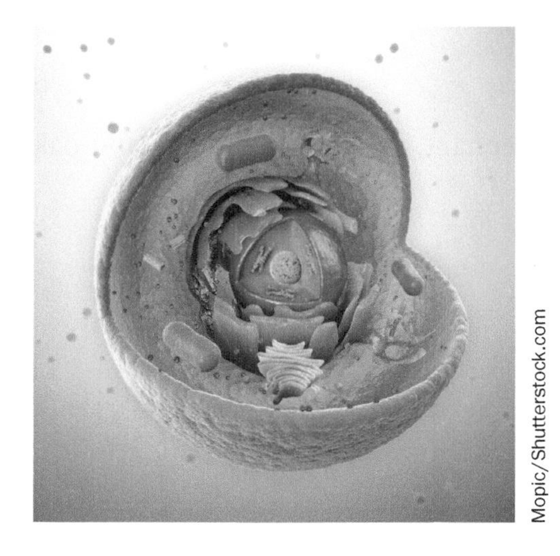
Mopic/Shutterstock.com

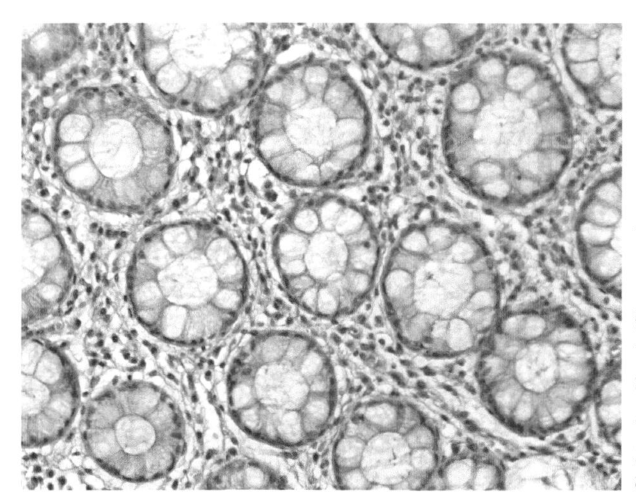
Christopher Meade/Shutterstock.com

Multiple tissue types with similar functions form an **organ**. The stomach and large intestines are examples of organs.

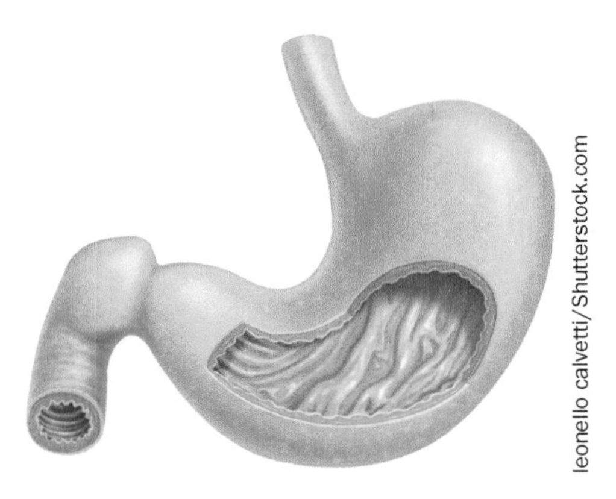

leonello calvetti/Shutterstock.com

Multiple organs collectively make up an **organ system**. The digestive system is an organ system made of the stomach, large intestines, among other organs.

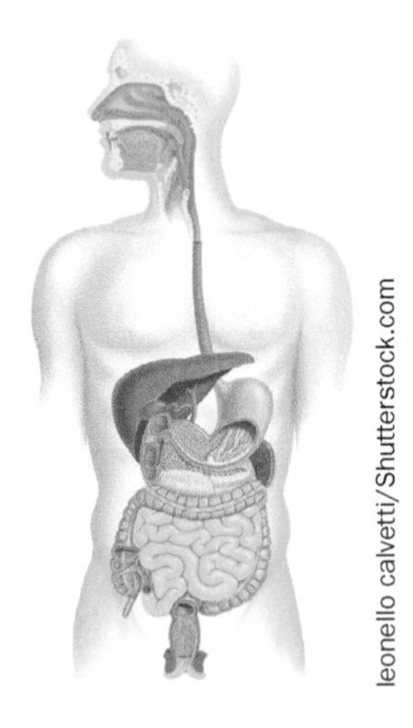

leonello calvetti/Shutterstock.com

Multiple organ systems sustain an **organism**. An elephant, mushroom, giraffe, and oak tree are examples of organisms.

Kletr/Shutterstock.com

A group of individuals of the same species living in proximity and interacting with one another is a **population**.

Kletr/Shutterstock.com

Multiple populations interacting in a given area is a **community**.

Kletr/Shutterstock.com

A community given its **biotic** and **abiotic** factors is an **ecosystem**.

Dennis Wegewijs/Shutterstock.com

The entire surface of earth given living and nonliving factors is the **biosphere**.

Alex Staroseltsev/Shutterstock.com

And there we have it—life from its smallest components to its largest.

History of Environmental Philosophy

eltoro69/Shutterstock.com

Ancient Greek philosophers such as Plato and Aristotle believed in the idea of **Provident Ecology**. This is the assertion that nature has an innate ability to maintain balance, so that any stimulus taking the earth away from its fulcrum would be counteracted with negative feedback. Under this assumption, if a species become overpopulated, thus threatening other species, nature would respond with an event such as a disease that would decimate enough of the overpopulated species to protect itself and its biodiversity. With this root assumption of the environment, extinction of any species, therefore, could not occur, since an outside force was overseeing its well-being. The Greeks did not see humans to be exempt from this system. They feared that by altering the earth through deforestation or agriculture, the population would face some type of outside divine retribution for its actions.

Fast forward now to the mid-19th century when, in 1859, Charles Darwin published *The Origins of Species* which examined **evolution** through the lens of **natural selection**. Darwin's book illustrated that among the innumerable variations in a population, some individuals' variations are better suited for the environment than others. Individuals better suited for the environment outcompete the others for resources and leave the most offspring. The concepts of evolution and natural selection will be discussed in greater detail in Chapter 3.

Charles Darwin.

Nicku/Shutterstock.com

The Greeks' ideas of Provident Ecology may seem oversimplified and contrary to the harsh realities of Charles Darwin's survival of the fittest concept of natural selection. But the two may not necessarily be mutually exclusive nor may they automatically cancel each other out. Let us say for example that the rodent species became overpopulated to the detriment of other species and that a small mammal species was well-adapted to hunting this small over-populated rodent species and began using it as a main food source. This scenario would contain the rodent population while increasing the overall fitness of the small mammals.

Furthermore, having 'high fitness' is sometimes the ability to cooperate with other species or share a home or resource and avoid direct competition. We will examine some of these examples in Chapter 5.

The early 19th century brought the Transcendentalist movement, a progressive and reformist movement which sprung from the writings of Ralph Waldo Emerson. The movement brought us such great philosophers and writers as Henry David Thoreau who wrote the classic book *Walden*, which described the intimate relationship between man and nature. Thoreau, while pointing out the correlation of natural phenomena to human societal customs, recognized the threat that industrialization posed to nature. He is thought of by many as the father of the environmental movement. Watching his native Concord, Massachusetts town being stripped of its natural resources, Thoreau wrote, "Each town should have a park, or rather a primitive forest, of 500 or a thousand acres, where a stick should never be cut for fuel, a common possession forever, for instruction and recreation."[ii]

The Native Americans for the most part defined themselves as being part of the landscape and viewed the physical or natural world and spiritual world as one unit. They acted in a manner to appease their 'spirits', and believed in living in a way that fit with nature as opposed to claiming ownership of it. Emulating this approach in our own lifestyles could do much to preserve out natural environment.

Whatever one's philosophy, throughout history there has always been incentive by humans to explore and assess their environment. Reasons are diverse and include: becoming familiar with the life cycles of crop-destroying pests such as caterpillars and aphids; knowing where to find quarry such as deer and fish; how to control a population of disease-causing vectors like mosquitoes; and even knowing the best crops to grow during a long drought or on marginal soil.

A Brief History of Conservation in the United States*

Compared to other scientific disciplines, conservation biology is still in its relative infancy, only emerging formally as a science in the 1980s. However, the beginning of the conservation movement in the United States dates back to the 19th century, largely in response to the rapid and unprecedented exploitation of natural resources associated with the post-Civil War era. The most conspicuous of these being the extensive, and largely unregulated, clearing of the Eastern forests for timber. As recognition of human-driven environmental impacts grew, an array of Western naturalists, scientists, theologians, and writers began to regard protection of the natural world as both necessary and a moral obligation. The emergence of a 19th century conservation ethic was also being driven by the growing belief that there were fundamental spiritual connections between humans and nature. Rather than nature being seen from a purely self-serving and utilitarian perspective, by the mid-1800s nature was starting to be seen as having both intrinsic and spiritual value worthy of protection and preservation.

The foundations of this spiritual philosophy can be traced back to the Greeks and what would come to be known as the Arcadian Vision. Arcadia, a province in what today is modern Greece, was the home of Pan, god of forests and watcher over shepherds. The inhabitants of Arcadia lived a simple pastoral existence and Arcadia itself became associated with the utopian view of humans living in perfect harmony with nature. Far from an abstract concept, the Arcadian Vision would be adopted as a central and persistent theme in the works of American authors such as Ralph Waldo Emerson (Nature[iii], 1836) and Henry David Thorough (Walden[iv], 1854). Indeed, literature would play one of the most significant and central roles in the development of a conservation ethic in the United States. Writers like Emerson and Thoreau would go on to refine the Arcadian Vision into the more formal philosophy of Romantic Transcendentalism and argue that nature should not be used solely for the extraction of its resources for material gain, but for the appreciation of its intrinsic, aesthetic, and spiritual value. Moreover, they believed that nature had the power to transform the human spirit and a place where one could draw nearer to, and commune with, God. This spiritual connection to nature reinforced the belief that humans should strive to live harmoniously with nature and that it should be valued for its own sake, not viewed solely for consumptive or utilitarian purposes.

The eminent naturalist, John Muir, echoed these sentiments and went on to become one of the most influential and effective preservationists in American history. Muir also believed that the preservation of nature was a moral obligation and became a leading advocate for protecting

*Contributed by Graeme Gissing.

nature so that it could serve a higher purpose, as a place for spiritual worship. Seminal to Muir's life-long commitment to conservation was the overwhelming sense of inspiration he felt upon arriving in the Sierra Nevada mountains in 1868. Awestruck by the sheer beauty of the landscape, this would soon evolve into a fierce dedication and advocacy for the preservation of wilderness in its natural and undisturbed state. A fan of both Emerson's and Thoreau's work, Muir was also a compelling writer but was able to successfully translate abstract romantic views into effective appeals designed to permanently protect wilderness areas in the form of national parks. For example, in 1892 he co-founded the Sierra Club with the mission of preserving the Yosemite Valley and other wilderness areas in the Sierra Nevada Mountains. As much a pragmatist as romantic, Muir also realized that he would not only need the support of the public, but the federal government. In 1901, he would find a friend and ally in the Whitehouse, President Theodore Roosevelt. Roosevelt, an avid bird watcher and outdoorsperson, supported Muir's preservationist goals and together they would prove instrumental in protecting natural areas in the West. When Muir took President Roosevelt on a camping trip to Yosemite in 1903, he was able to convince Roosevelt of the importance of preserving more of the Sierra Nevada as federal land. Yosemite National Park as we know it today took shape in 1906, when Roosevelt took back control of the Yosemite Valley from the state of California and protected the entire region as Yosemite National Park. Roosevelt would continue his ambitious conservation agenda, establishing 51 wildlife reserves which would later become the National Wildlife Refuge System of the U.S. Fish and Wildlife Service, creating 18 national monuments, and five national parks by the end of his administration.

Although great strides were being made by preservationists such as John Muir, an opposing conservation philosophy was emerging based on a utilitarian view of resource management developed earlier in Europe. Known as the Resource Conservation Ethic, it differed from the preservationist view in that resources were to be utilized but managed using scientific principles. One of the most influential proponents of this Ethic was Gifford Pinchot who took the position that conservation should not mean preserving nature, but the wise and efficient use of natural resources in manner that was informed by science. As a young man, Pinchot had witnessed the devastating effects that unrestricted logging had on the Pennsylvania landscape and realized that this unsustainable practice was unlikely to the meet society's needs over the long run. Although holding a fundamentally different view than the preservationists, Pinchot is now considered one of the giants of American conservation because of his efforts to protect America's forests. Although Pinchot's view of nature was primarily as a source of exploitable natural resources, he also believed that such resources should be used for societal good and, more importantly, conserved for future generations. This led him and others to advance the now familiar concept of maximum sustained yield, where a constant level of resource extraction equals the rate of renewal of a resource such as timber[v]. In a sense, the Resource Conservation Ethic was a pragmatic management-oriented approach to the ever-increasing demand for natural resources with Pinchot defining conservation as "the foresighted utilization, preservation and/or renewal of forests, waters, lands and minerals for the greatest good of the greatest number for the longest time"[vi]. Appointed in 1906 as the first chief of the newly formed U.S. Forest Service, Pinchot undoubtedly played a major role in shaping American forest conservation policy by institutionalizing the view that timber was an exploitable resource but should be harvested sustainably using the careful application of scientifically sound management practices. This conservation strategy established a deep divide between the Romantic Transcendentalism (preservationist) and Resource Conservation (utilitarian) schools of thought which persists to this day.

One of the limitations of the Resource Conservation Ethic is inherent to its central theme, that natural resources are primarily to be utilized for human consumption, even if done sustainably. One of the most influential critics of this approach was Aldo Leopold. Leopold saw nature as a collection of interconnected systems rather than a storehouse of commodities with humans being members, rather than conquerors, of the larger biotic community. For Leopold and others, this translated into a new definition of conservation's goal, away from sustaining commodities toward a more complex goal of sustaining healthy ecosystems[vii]. This premise became the foundation of Leopold's ecocentric view, which asserts that the entity to which both values and rights apply is the biotic community, not any individual species. In other words, he believed that true conservation and effective land management requires an overriding ecological perspective. Leopold's most famous work, A Sand County Almanac[viii] of 1949, proposed a Land Ethic which made the ethical treatment of land and resources a central issue in conservation. Today, many conservation biologists see themselves as heirs to Leopold's legacy in attempting to restore ethics and values to the science of conservation. Undoubtedly, contemporary conservation biology draws upon areas ascribed to each of these three founding Ethics.

This brief history of conservation in the United States illustrates clear divisions in both philosophy and practice. Driven largely by the work of Michael Soulé and colleagues, the formalization of conservation biology as a science in the 1980s would intentionally attempt to bridge these divides. This was in evidence at a seminal and intentionally multidisciplinary meeting held at the San Diego Wild Animal Park in 1978 that became known as the First International Conference on Conservation Biology. Followed closely by Soulé and Wilcox's landmark publication, Conservation Biology: An Evolutionary-Ecological Perspective[ix] in 1980, the formation of the Society for Conservation Biology in 1985, and the creation of the journal Conservation Biology in 1987, the new scientific discipline of conservation biology had officially arrived. Although born out of the growing 20th century crisis of increasing extinctions, contemporary conservation biology focuses on understanding and evaluating the human-caused impacts on global biodiversity rather than the conservation of individual species. Soulé described this new science as a "crisis discipline" with the desired outcome predetermined -- the use of science to develop tools and strategies for preserving biodiversity and ensuring the persistence and viability of ecosystems[x].

Why Study Environmental Science?

It may seem as though every time you open a newspaper or turn on your computer news headlines warn us of escalating carbon emissions, rising global temperatures, droughts in countries accustomed to abundant annual rainfall, water shortages, or polar ice caps and permafrost melting and species decline. These cautionary news articles indicate that our overexploitation of the environment and excess waste have caught up to us in damaging and alarming ways.

Mavrick/Shutterstock.com

Many serious environmental problems today started with the industrial revolution which began approximately 200 years ago. During that time, factories were constructed all over the world and technology became widespread. The world observed a notable improvement in health care and sanitation, and as a result, the death rate decreased while the carrying capacity and life span in developing nations increased. Factoring in the increased birth rate and decreased death rate, the human population increased at an unprecedented rate.

The factories that were built used coal for fuel. Coal is a carbon-based fossil fuel. When coal is burned, carbon reacts with free oxygen in the atmosphere to create carbon dioxide, CO_2. CO_2 reacts with infrared (long-range) light and either emits light back into the upper atmosphere or sends it back to earth, causing an overall warming effect. We call this **global warming**. Any atmospheric gas that is linked to overall warming of the earth's lower atmosphere is called a **greenhouse gas**. Carbon dioxide is not the only greenhouse gas, but it is perhaps one of the most damaging. Other greenhouse gases are water vapor, methane, nitrous oxide, and ozone. According to NASA, earth's surface temperatures have increased .8 degrees Celsius since the year 1880. A small increase may seem insignificant but its impact on resources and biodiversity is devastating. Climate change and its effects will be explored in Chapter 13.

How Do Scientists Evaluate the Environment?

Science is an **iterative process**—that is, it is repetitive and built on itself from precious research, experiments, and data. Science is based on observation, testable theory, peer review, and empirical data. All advancements in science are done so through the **scientific method**—a series of key steps to scientific investigation and inquiry. This process is based on *deductive reasoning* which involves ruling out various possibilities through observation and empirical evidence and noting what remains or cannot be ruled out through the deductive process.

The experimental process is peer-reviewed, which contributes to the legitimacy of the process. Experts in the field oversee the entire experimental process to make sure it adheres to scientific standards.

The scientific method begins with our **observations** and conceptions about nature. For instance, you may notice that there is an observable lack of biodiversity on the rocky shores of polluted beaches—much less than that of pristine beaches with high air and water quality.

Since the scientific method is a process that builds on itself from other data and research, one would look into similar research conducted at other universities or institutions regarding biodiversity of aquatic habitats, and more specifically, rocky shores or intertidal zones. Their methodology and conclusions would be used as inspiration and guidelines for new scientific investigation.

The next step is **hypothesis** generation. The hypothesis is generally a statement or or a testable theory that often set out to be falsified, or disproved, through data collection and observation. In our above example, a suitable hypothesis may be "There is no correlation between pollution and the biodiversity of rocky shores."

The reason for this is that you must have empirical data to back up a statement or a hypothesis. It is impossible, for instance, to have complete empirical data to support the statement, "Cleaner intertidal zones harbor more biodiversity than intertidal zones with higher pollution." One must account for all beaches, past, present and future. It is more correct, therefore, to state, "There is no correlation between pollution and the biodiversity of rocky shores." This way, any evidence to the contrary will disprove the hypothesis.

The next step is the **experiment** portion in which data are collected. The experiment is based on the **variable**, the cause for the experiment as well as the factor being adjusted and manipulated. In our example, the variable would be the varying amount of pollution. The experiment is compared to a **control**—the replicated experiment *without* the variable (a pristine rocky shore with no pollution for instance). The process must be repeatable, meaning it is to be performed more than once with similar results.

Once the experiment is complete, a conclusion is reached in which the hypothesis is either accepted or rejected. If a hypothesis cannot be disproved after multiple studies, it becomes a **scientific theory**. If a scientific theory cannot be disproved, it becomes a **scientific law** (the law of gravity, for example). The scientific method has been used to conduct research on climate change. (see Chapter 13) as well as other environmental concerns, which we will examine in the following segment.

I hope you can appreciate the integrity of the scientific method and the importance of scientific literacy. Our lives are predicated on science; from health, to GMOs, to vaccines, to the environment, to nuclear energy, to nutrition, etc. It is important that we are able to disseminate scientific literature in order to do our due diligence and become our own advocate. Scientific investigation teaches us to think critically, and critical thinking is the cornerstone of problem solving and independent thought.

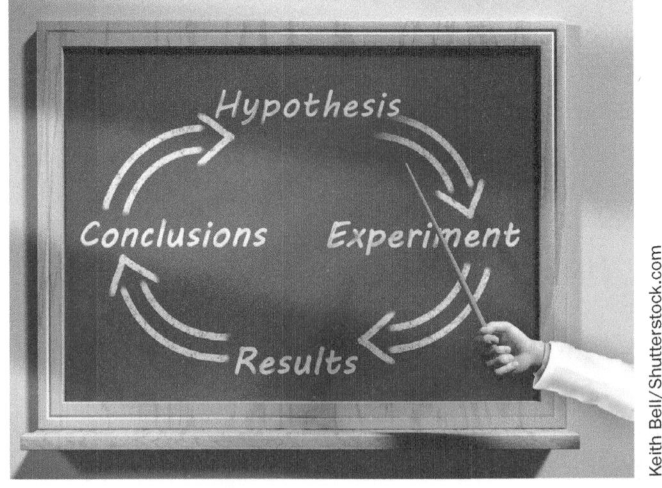

Keith Bell/Shutterstock.com

The **IPCC** (Intergovernmental Panel on Climate Change) is a Nobel Peace Prize-winning panel overseen by the United Nations whose mission is to examine and assess climate change and its impacts through a scientific lens. Scientists participate *pro bono* on a volunteer basis and must adhere to the strict guidelines of the scientific method as well as the standards of the IPCC. Examples of such current research include the scientific examination of ice cores, soil, sediment, satellite images, as well as land and water temperatures.

Major Causes of Environmental Problems

Most of our environmental problems can be traced back to **overpopulation**. According to census.gov, in early 2019 the global population was over 7.5 billion people. According to the United Nations Population Division (DESA) the global TFR is roughly 2.5. The earth yields a finite supply of resources, most of which are nonrenewable (land, freshwater, and minerals, for example) and may take hundreds, if not, thousands or millions of years to regenerate. Such a large population creates a high demand for these resources, which will eventually exhaust them. There is not enough resource supply to indefinitely sustain our growing population. Furthermore, some of the methods used to harvest resources are done unsustainably, causing irreversible damage. Unchecked human population growth also leads to pestilence, poverty (another cause for environmental problems—see paragraph that follows), and droughts, pollution, contaminated water, desertification, among many other problems associated with environmental issues and issues concerning human health.

spirit of america/Shutterstock.com

Another cause for environmental problems is **overexploitation**, meaning that we take goods from the earth faster than they can be replenished naturally. Overfishing, overgrazing, overmining, and overlogging are examples of overexploitation. Some of the resources we take from the earth are renewable, but even so, the rate at which we use them and methods of obtaining them can lead to extinction of various species or a dramatic

psv/Shutterstock.com

loss of biodiversity. It is difficult for ecosystems to recover from extreme biodiversity loss. (We will be examining some of these resources in later chapters.)

dominique landau/Shutterstock.com

Poverty is another major cause for environmental problems. Poverty often leads to poor health care, overpopulation, and disease. Weak environmental policy may lead to some people resorting to environmentally damaging practices such as unregulated disposal of hazardous material, burning rubbish (releasing harmful chemicals into the atmosphere), or **poaching** (illegal selling or slaughtering animals for tusks, fur, or other valuable body parts). Poaching occurs when animal parts are worth a lot of money. Countries with little governance may be powerless to stop or enforce laws against poaching or exploitation of natural resources.

Furthermore, wealthy companies that remove, harvest, or destroy plant and wildlife in areas rich with biodiversity (such as the tropics) do not adequately compensate the host country or its citizens. Since biodiversity is a type of currency and worth a great deal of money (in addition to other immeasurable value); this results in a net loss of capital and/or money for the host country, perpetuating a cycle of poverty. This may be curbed if companies would compensate host countries and their citizens appropriately or if host countries were encouraged to use their natural resources sustainably.

Affluence is our final major cause for environmental problems. Affluence is a relative term, as the average westerner may appear to have considerable wealth compared to the average person of the developing world. In the context in which we are using the term, affluence leads to excess consumption and excess waste. It often means excessive air travel on planes, private jets, and large luxury vehicles. These all use a great deal of fuel and leave a large carbon footprint. Wealth often contributes to diets rich in meat and dairy which have big carbon, land and water footprints. Wealth may also lead to an overconsumption of gadgets and electronics that leave excess toxic and nondegradable waste and increase pollution. It can also lead to overdevelopment of land for large homes and shopping centers. Overdevelopment disrupts species' habitats, compromises biodiversity, and generates waste.

Will History Keep Repeating Itself?

The fate of Easter Island, a small remote Island located in the Pacific Ocean, remains not only a mystery but also a cautionary tale. It is widely accepted that their inhabitants engaged in the typical slash and burn agriculture, deforestation, and overexploitation of resources that ultimately may have led to their demise.

Desperate for solutions as they tackled poor quality soil and lack of soil nutrients, Easter Island inhabitants resorted to a means intended to potentially bring nutrients back to their soil. They created what one might refer to as a "stone garden," in which they crushed stones into smaller bits and scattered them onto fields, creating an uneven landscape of rock. As wind and elements moved over this choppy surface, an uneven wind current was created, releasing minerals from rock (minerals are an important component of soil - please see Chapter 8), bringing soil back to a healthy and nutrient-rich composition. This may likely have facilitated growth of vegetation.[xi]

As the scenario of Easter Island inhabitants suggests, environmental issues and concerns, such as the lack of available resources and poor quality soil, are not only a 21st century dilemma.

In the 1800s England experienced a conundrum when faced with a dramatic loss in wood fuel supply as a result of deforestation. Their wood was used to build ships, temples, and edifices and for fuel. As a solution they turned to kelp, a type of large multicellular algae. Kelp was collected, dried, burned, and eventually used for many functions including fuel, paper, glass, and many more.

Ancient Greeks also faced a similar deforestation problem and were forced to improvise with available resources. As we read in previous paragraphs, Henry Thoreau expressed grave concerned for deforestation in his book *Walden*.

Growing concern over the environment also brought us clever solutions—creating stone gardens and resorting to kelp as a wood alternative—are just a couple of examples in our vast human history that demonstrate how various civilizations, facing diminishing natural resources, have integrated natural resources in an advantageous manner to solve environmental problems.

Please take a moment to reflect upon our culturally dominant philosophies surrounding the environment. Do we see the environment as a potential source of revenue and/or a business commodity? Or do we see it as a unified working ecological system in which we are part? The reason that it is essential to examine our core beliefs about nature is that these core beliefs inform our actions and how committed we are to saving our earth.

Will humans learn from our past and acknowledge the limitations of earth's natural resources? Will we acknowledge our place in the biotic world, understanding that we are an extension of the environment, not a separate entity? Will sustainable use of resources and protection of wildlife and biodiversity become the norm? Will human ingenuity present us with alternatives to solve many of our present day environmental problems? Will those living with affluence learn to curb their consumptive lifestyle? Will those living with abundance assist those in need and help break the cycle of poverty? And most importantly, how can each one of us be part of the solution, rather than a part of the problem?

Review Questions

1. What is the difference between a carbon and an ecological footprint? Why is the scientific method considered iterative.

2. Define "abiotic" and "biotic." Please provide an example of each.

3. What are the major steps of the scientific method? Why do we set out to prove the hypothesis wrong?

4. What are the major threats to causes of environmental problems?

5. What is the difference between the chemical and the physical environment? Provide examples of each.

6. Why is Henry David Thoreau often considered the "father" of environmentalism?

Discussion Questions for Class

A. Santa Monica, Portland, Seattle, Vancouver, and several European metropolitan areas are strong examples of green cities. Discuss how they have achieved this moniker. What types of municipal transportation is available? Are there bike paths and roof gardens?

B. How green is your city? Do you have native gardens and recycle gray water? Do you have bike paths? Does your city use natural gas or electric vehicles? Is recycling readily available? How could it improve?

C. Are we headed toward a greening of our society in which we will be given incentive to live more sustainably?

D. Is it our civic duty and personal duty live more sustainably?

E. Should households with larger ecological footprints be taxed or be required to pay restitution? Or should lowering footprints be more incentivized? (Tax breaks, etc.)

F. Is it possible that anthropogenic causes may have cataclysmic effects similar to the volcano eruption in Iceland in the 1700s?

G. Why is it necessary for the scientific method to be interactive and peer-reviewed?

H. Should companies be taxed as well as individual households for water, carbon, and pollution? Should there be exceptions? What incentives should be offered?

I. Can you think of any other major causes for environmental problems or a solution to thee problems? Please share your ideas.

Find an environmental non-profit organization that speaks to you or supports a cause that you are interested in. Get involved in local campaigns. Find out how you can volunteer. Think of a household product you use every day. It may be a toothpaste or laundry detergent. Note the company or corporation that manufactures this product you use often. Do some background research on this company and its product(s). How does this company's product(s) affect the environment? If its products have an adverse effect, stop using the product and switch to a more ecofriendly brand. Share your knowledge on your social networking sites.

Recycle and buy vintage clothing or consignment.

When ordering takeout, bring your own containers, and bring recyclable bags to the grocery store.

Find out of there are any county/state/Federal parks near you doing restoration projects. Pursue any volunteer possibilities.

Carry your own portable fork/knife/spoon kit.

Use detergents with natural jasmine oils (you can purchase these at Trader Joe's) and with less phosphorus, and use a French press coffee maker (no throwaway filters).

Read labels of snacks, candies and soaps. Avoid any product that contains palm oil or its derivatives.

Find environmentally friendly products with natural ingredients and recycled or compostable packaging.

Line-dry clothes instead of using a dryer.

List your own suggestions here.

Endnotes

i. Greg Neale, The Guardian, Thursday 15 April 2010, 13.11 EDT.
ii. *The Writings of Henry David Thoreau* (Boston: Houghton Mifflin, 1906)
iii. Emerson, R. W. Nature. 1836. In: The Norton Anthology of American Literature. Vol. 1. (eds. Nina Baym et.al. New York: W.W. Norton & Company, 1989, 903–31.
iv. Thoreau, H. D. (1854). Walden. Boston: Houghton, Mifflin.
v. Van Dyke, F. (2008). Conservation biology: foundations, concepts, applications. Springer Science & Business Media, 1–26.
vi. Pinchot, G. (1910). The fight for conservation. Doubleday, Page, and Company, New York.
vii. Meine, C. (2010). Conservation biology: past and present. In Conservation biology for all (eds. N. S. Sodhi, P. R. Ehrlich), 7–26.
viii. Leopold, A. (1949). A Sand County Almanac. New York: Ballantine Books.
ix. Soulé, M. E. and Wilcox, B. A., eds (1980). Conservation biology: an evolutionary-ecological perspective. Sinauer, Sunderland, Massachusetts.
x. Soulé, M. E. (1985). What is conservation biology?. BioScience, 35(11), 727–734.
xi. (From J.B. Mackinnon, "The Once and Future World.")

Chapter 2

A Scientific Look at Life, Atoms, Molecules, Matter, pH, and Energy

DID YOU KNOW?

Coal is a type of sedimentary rock with a very high carbon content; between 40% and 90% carbon by weight. It forms at the end of a prolonged process in which dead plant matter undergoes many geological changes. As coal gets older, it becomes harder and blacker and contains a higher carbon concentration. It starts this process initially as peat, then becomes young coal, lignite, bituminous (the most abundant in the United States), then anthracite, the hardest coal of all. Coal can be formed in freshwater or saltwater areas. Coal with a high sulfur content most likely formed in the presence of sea water. Due to its ample supply, high net energy yield, and low cost, coal has traditionally provided the major energy source for electricity in the United States. Coal is made up of many carbon atoms that come together by collectively bonding and forming a larger solid substance. When ignited, coal releases large amounts of energy (in the form of heat) as its carbon atoms are released from these bonds and dispersed into the atmosphere. Once dispersed, they react with atmospheric oxygen. As bonds are broken and reformed, the result is carbon dioxide gas, often written in its **molecular formula** CO_2. Carbon dioxide is a **molecule**, a group of atoms, in this case carbon and oxygen, bonded together to form a larger group of molecules. Collectively, the larger group of molecules has its own unique properties that are different from those atoms of which it is comprised. Carbon dioxide, a greenhouse gas, has been linked to increasing global temperatures. It reacts with long-range light, trapping it and causing an overall increase in the amount of heat in the earth's atmosphere. This increase in heat causes molecules to gain momentum and vibrate at a faster rate (we experience this as an overall temperature increase in the earth's lower atmosphere). The reason for this increase in momentum can be explained by the fact that heat is a type of energy. **Energy** is defined as the capacity to do work. There are two main types of energy, potential and kinetic, that will be discussed later on in this chapter.

Chemistry of Life

In order to explore environmental science, we must first examine the building blocks of life, and the microscopic materials that make up the living world. We will start with the **atom** (the smallest unit of matter that retains its own chemical properties), and later in the chapter, work our way up the ladder of life toward the biosphere—the entire surface of earth and its living components.

Petr Vaclavek/Shutterstock.com

The nucleus of an atom contains both **protons** (positively charged particles) and **neutrons** (particles having no charge.)

The unique chemical properties of an atom are determined by its **atomic number**, which is the number of **protons** in the nucleus.

The **atomic mass** of the atom is the number of protons plus the number of neutrons measured in daltons.

Electrons are negatively charged dynamic particles that have a high probability of circulating around the nucleus of the atom. There is a high probability of having the same number of electrons as there are protons found in the nucleus. When two atoms come into close proximity, they may share electrons, forming a **covalent bond** and becoming part of a larger **molecule**.

Ions are atoms or molecules with either a positive or a negative charge. If it has a positive charge, then it has lost an electron(s), giving a net *positive* charge. If it has a negative charge,

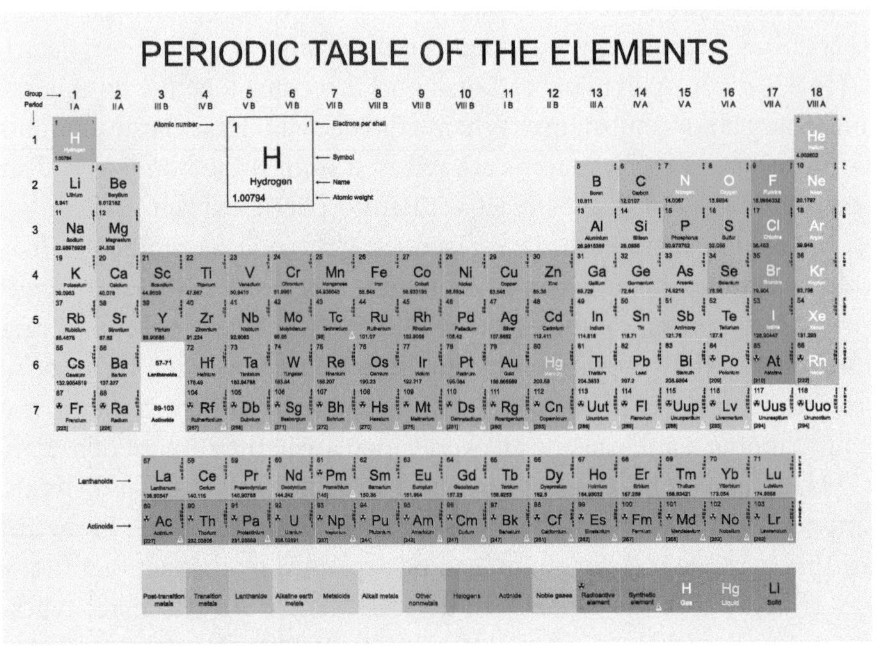

concept w/ Shutterstock.com

then the atom has gained an electron(s), giving it a net *negative* charge. A **cation** is a positively charged ion; an **anion** is a negatively charged ion. Since opposites attract, an **ionic bond** is a bond created by the attraction of a negatively charged anion to a positively charged cation. Table salt, NaCl (sodium chloride), is formed through ionic bonds. The positively charged Na^+ ion bonds with the negatively charged Cl^- ion to make salt!

An **element** is the smallest unit of matter (made up of atoms) that retains its own unique properties and cannot be broken down by conventional physical means (heating, distillation, etc.). Carbon, hydrogen, sulfur, nitrogen, helium, neon, and boron are all examples of elements. You may be familiar with the periodic table that is an index of all the elements.

Compounds are molecules that are made up of different elements always found in a consistent ratio. H_2O, for example, is a molecule that always contains two hydrogen elements to one oxygen element.

Isotopes are molecules whose atoms have varying numbers of neutrons. Isotopes are often unstable and decay, sometimes becoming radioactive. The term *radioactive* may seem ominous but, in fact, radioactive isotopes can have a wide variety of beneficial uses. Their functions range from the medical use of checking for ulcers and bone abnormalities to archaeological uses such as determining the age of an ancient fossil. It is when isotopes are extremely radioactive that they are harmful, causing mutations in DNA, burns, and skin ulcers. Elements that become dangerously radioactive tend to have a high molecular weight (like Uranium).

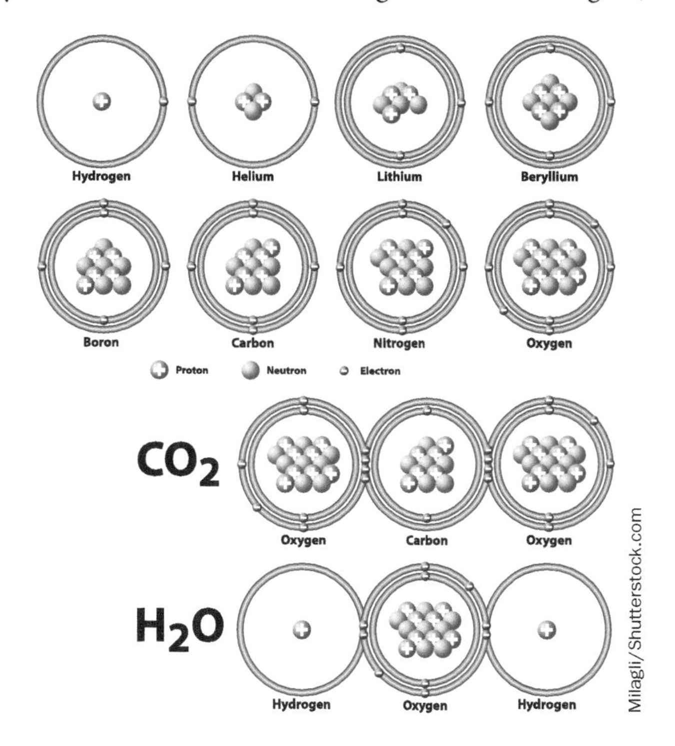

Milagli/Shutterstock.com

A *chemical change* is when one substance transforms into another. Oxidation or rusting is an example. A *physical change* is when the chemical properties remain constant but its form transforms. Chemical changes can change chemical properties such a pH and reactivity with water.

Living things are made up of **organic molecules**—that is, molecules made up mostly of carbon and hydrogen. Organic molecules contain other elements as well. We can simplify this by saying that living things are mostly **CHNOPS**, an acronym for carbon, hydrogen, nitrogen, oxygen, phosphorus, and sulfur, the most abundant elements found in living things and the molecules of life.

There are four major classes of organic molecules that are universal to all life:

1. Proteins,
2. Carbohydrates,
3. Lipids, and
4. Nucleic Acids

Proteins are comprised of 20 different building blocks called **amino acids.** (I am sure you have, at some point, heard the term "essential amino acid." The body does not synthesize essential amino acids so they must be taken in from the diet and incorporated into the body). Proteins are among the most diverse class of organic molecules due to their variety of functions. They compose muscle, fingernails, they are a major component in skin, hair, they make up hemoglobin, they are **enzymes** (biological catalysts), and antibodies (an integral part of our immune response), to name only a few.

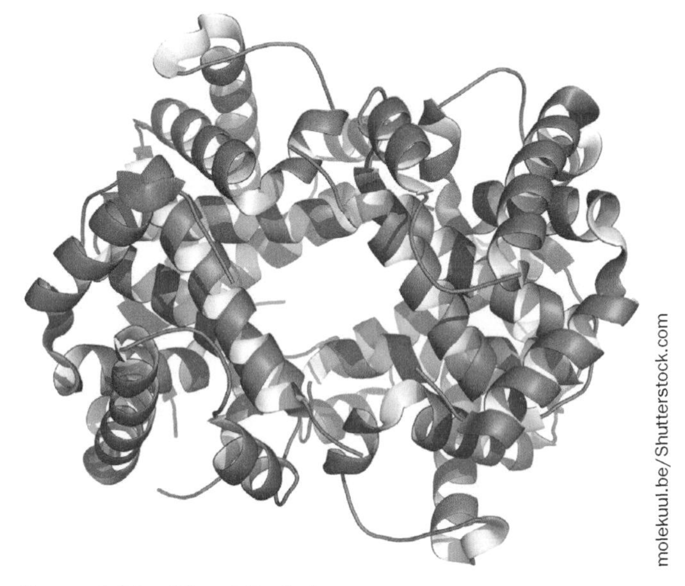

molekuul.be/Shutterstock.com

Hemoglobin; Blood Protein

"Carbohydrate" is another name for "sugar." Carbohydrates have a chemical formula of CH_2O. Carbohydrates are molecules that store energy. They are called **glycogen** when stored in animals and **starch** when stored in plants. As humans, we possess the enzymes that can break down starch and we then use it as an energy source. Foods rich in starch include potatoes, pasta, beets, turnips, rice, and wheat. **Cellulose** is a type of structural carbohydrate that makes up plant cell walls (it does not store energy). In terms of our diet, we are familiar with it as fiber. We do no not possess the enzymes to efficiently break down cellulose, but need dietary fiber to keep our digestive tracts healthy.

Lipids are fats and characterized by the fact that they are **hydrophobic**—meaning that they do not react with water. Fats have at times acquired a bad change reputation to connotation because we often associate them with adipose tissue, which swells when we consume too many calories and fat in our diet. Despite their negative connotation, fats are essential to the body. They compose the cell membrane, they store more energy than carbohydrates, and they are hormones such as testosterone and estrogen. Nonetheless, we should avoid a diet high in fatty acids.

And finally, nucleic acids such as DNA (DNA is an example of a nucleic acid), which acts as our genetic blueprint—the information that codes for our gene expression. Our DNA is stored in the nucleus of the cell. With the exception of identical twins, everybody's DNA is unique.

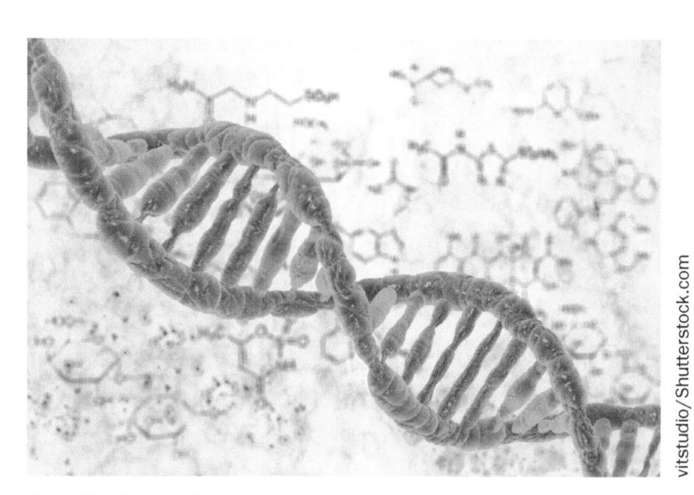

A DNA Molecule

vitstudio/Shutterstock.com

Trace elements are elements, such as iron and magnesium, which are not part of CHNOPS but are needed by the body in small doses. Some examples of these micronutrients are Zinc, Chromium, Manganese, Copper, Iodine, and Selenium. Their functions vary from helping carry oxygen throughout the body to ensuring healthy endocrine function, to helping with brain and muscle function.

Small doses of trace elements are essential in humans' and animals' overall health. A selenium deficiency in livestock, for example, can make them more susceptible to disease and parasites.[i] Iodine in humans is essential for healthy endocrine function. A person who lacks sufficient Iodine in their diet may develop goiter, an enlarged thyroid. But too much of these micronutrients can sometimes be harmful or toxic to the body.

Inorganic molecules are chemicals that are not carbon based. They can occur naturally and we may ingest them for dietary needs; examples include iron, zinc, selenium, iodine, and magnesium. They also include calcium and phosphorous that are needed for bones, shells, and antlers.

But inorganic molecules are also often synthetic, meaning they are created in a laboratory. Inorganic chemicals that are synthetic are potentially harmful to the body because the body does not recognize them and cannot break them down properly. You might say that they "speak a different language" than your body. Many industrial chemicals, fertilizers, herbicides, and pesticides are inorganic. They cause many potentially harmful side effects to our health and the health of wildlife and the overall environment.

Fun Fact

Pre World War II, many of the insecticides used by humans came from secondary volatile plant compounds (menthol and citronellol, for example, which are organic). Although effective, these plant-derived compounds evaporate quickly and tend to lose their efficacy upon evaporation. Determined to find a better solution, the U.S. Army developed the inorganic chemical DEET, an effective insecticide that was very effective at thwarting bites from mosquitos and ticks. (DEET does not make it hard for insects to bite humans, it makes it difficult for them to smell or detect humans.) DEET became available to the general public in the 1950s and is still used today (and the EPA supports its use and safety present day).

Although very effective as an insect repellant, soon after its introduction, some of the unpleasant side effects and negative reactions from this chemical became apparent. This in part is because DEET is a chemical that is not from natural origins and is not recognized by the human body.

Deciding that it is not worth the risk, many people and pharmacists are moving back toward organic insecticides derived from secondary volatile plant compounds. In fact, many people are more comfortable with medicines: emollients, antimicrobials, bioactive compounds, and insecticides that have natural origins. This is reflected in the growing demand for naturopathic medicine, which tends to avoid traditional pharmaceuticals.

Properties of Water and pH

Water (H_2O) is a ubiquitous **polar** molecule with many unique properties and is the only substance that is commonly found in solid (ice), liquid (water), and gas (water vapor) form. Polar molecules have two distinct sides each with distinct charges (positive or negative). In water molecules, the oxygen end of the molecule carries a net negative charge while the hydrogen-end carries a net positive charge.

Since opposite charges attract one another, water creates bonds with itself; therefore, water is considered to be **cohesive**. Water also bonds with any substance carrying either a positive or negative charge, so water is also **adhesive**.

Because water molecules have a strong affinity for each other, the **hydrogen bonds** they create are strong, absorbing a lot of heat without breaking apart. As a result of this, water has a **high specific heat**, meaning it takes a lot of energy to raise one cubic centimeter of water to 1°C. You may notice this while you are waiting a long time for your water to boil! Since living things are roughly 60% water or more, water's high specific heat enables us to maintain

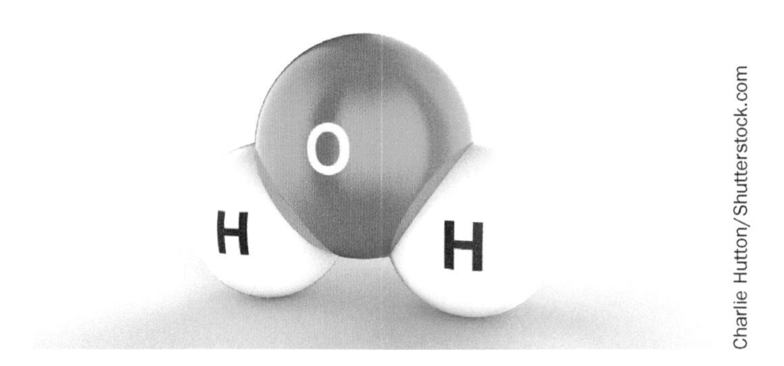

Charlie Hutton/Shutterstock.com

homeostasis (a consistent internal environment), a factor essential to sustaining life. Mild night-time temperatures of coastal areas can also be attributed to water's high specific heat. Oceans absorb a lot of heat during the day then release it at night, having an overall temperate effect on coastal temperatures while inland temperatures may plummet.

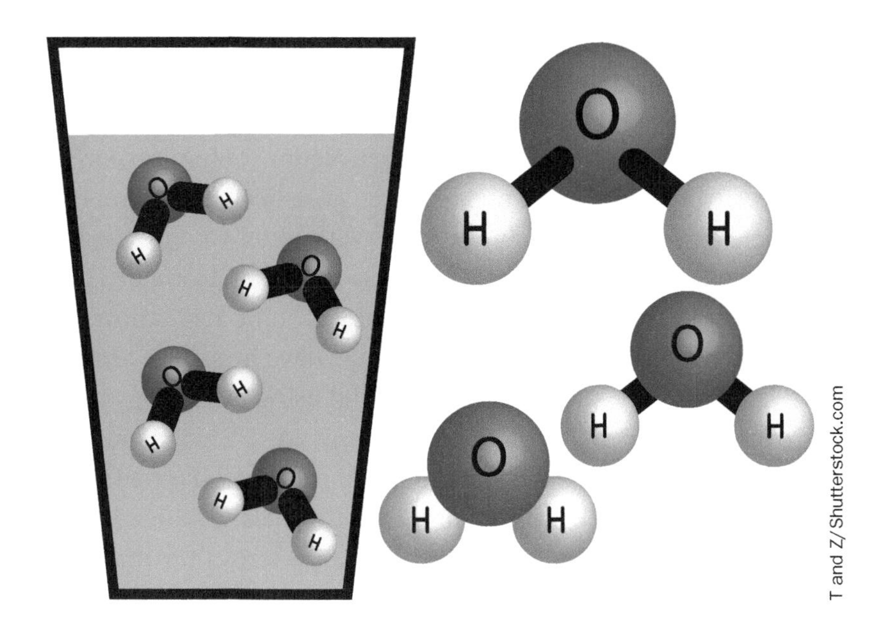

T and Z/Shutterstock.com

pH is a measure of how acidic or basic an aqueous solution measures. When water (H_2O) disassociates, it breaks into equal parts of H^+ cations and OH^- anions. When equal concentrations of OH^- and H^+ ions are given off, as is the case with water, the substance is said to be **neutral** and has a pH of 7. If a substance gives off an excess of OH^- ions, it is a **basic** solution and has a pH above 7 (the pH scale runs from 1 to 14, with 1 being extremely acidic and 14 being extremely basic). Many common household cleaners such as bleach and oven cleaner are basic. If a solution gives off an excess of H^+ ions, then it is **acidic** and has a pH of less than 7. Highly acidic solutions such as battery acid are corrosive and damaging.

Your blood tissue is basic and acts as the body's **buffer**, a substance that helps maintain a consistent pH. Blood will increase pH if body fluid becomes too acidic and similarly lower pH if body fluid becomes too basic. In this sense, blood helps the body maintain homeostasis.

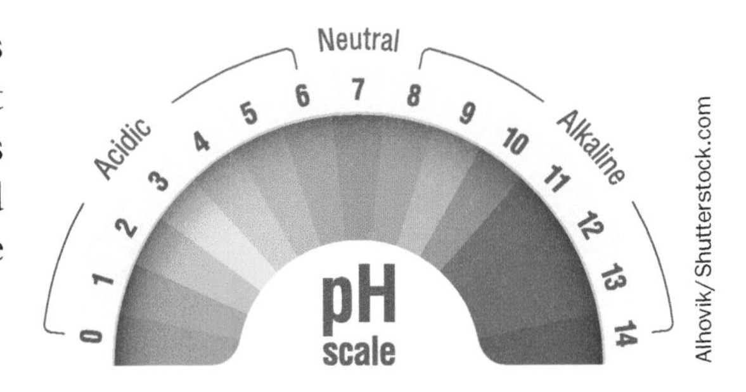

Alhovik/Shutterstock.com

Acid Rain

Acid rain is rainwater that has a pH of 5.7 or less. It is damaging to aquatic life, vegetation (it can lead to a weakened immune system in trees), buildings, statues, and infrastructures and soil. It can put nitrogen into the soil and eventually lead to eutrophication of our lakes, streams, and other bodies of water. Many living organisms are intolerant to low pH, so their enzymes denature as the pH decreases. Acid rain is a by-product of burning fossil fuels and is therefore another environmental problem whose origins can be traced back to the industrial revolution and our reliance on fossil fuels.

Both coal and **petroleum** are fossil fuels. This means they were created from the fossils of living organisms that perished millions of years ago. The remains of these living organisms have been exposed to excess heat and pressure over the years, forming coal and petroleum. Because the remains were once living organisms, coal and petroleum contain carbon as well as traces of sulfur and nitrogen (all CHNOPS elements). When coal is burned, carbon reacts with water in the atmosphere to make **carbonic acid**. Sulfur reacts with water vapor to make SO_2, sulfur dioxide. In the presence of other molecules, sulfur dioxide may react to create H_2SO_4, **sulfuric acid**. Nitrogen reacts with oxygen to make **nitrogen oxides**. Carbonic acid, sulfuric acid, and nitrogen oxide are all responsible for causing acid rain. You may have heard the term "clean coal" in the news or from environmentalists. This refers to the act of using technology to reduce the amount of sulfur and nitrogen in coal as well as limit carbon emissions. This reduces the amount of acid rain and other negative impacts that the combustion of fossil fuel has on the environment.

Carbon dioxide also eventually gets absorbed by the oceans where it reacts with water to form carbonic acid. As more carbon dioxide is released as a result from the use of fossil fuels, higher concentrations of carbonic acid are found in the oceans, lowering its pH and causing **acidification** of our oceans. This has catastrophic consequences for aquatic life whose **range of tolerance** for changes in pH is very narrow.

Energy

Energy, as described above, is defined as the capacity to do work. Energy is typically either **potential** or **kinetic**.

Potential energy is energy that is stored and ready to be released in some capacity. Examples of potential energy include the energy stored in a battery, the energy stored in the carbohydrates of bread (also considered **chemical energy**), or the energy stored in the springs of a diving board.

Kinetic energy on the other hand is energy *in motion*. Once the diver jumps and activates the spring, the energy has changed from potential to kinetic energy. Water flowing is an example of kinetic energy, as is the flow of electrons when a battery is used.

The following types of energy are typically associated with kinetic energy, but often have distinct definitions: heat or **thermal energy** occurs when temperature increases or decreases. Typically, molecules increase in movement with increase in temperature. In this case, the energy content of a system increases. **Electromagnetic energy** is a type of energy that is measured in waves (light is measured in wavelengths.) **Electrical energy** is a result of the flow of electrons.

According to **the first law of thermodynamics**, energy cannot be created or destroyed, it can only change form. According to the **second law of thermodynamics**, when energy changes form, most of the energy is lost as heat. To illustrate these concepts, think about the last time you exercised. When you go for a jog, your body temperature increases and you subsequently sweat to cool down and maintain homeostasis. Your body is accessing the potential energy stored in the covalent bonds of carbohydrates you ingested for energy. As you exercise, your body transforms this potential energy to kinetic energy, all the while your body temperature increases as the energy is lost as heat.

The second law is also illustrated in the use of an incandescent light bulb. Incandescent light bulbs have very low efficiency since 95% of their energy is lost as heat! This means that only 5% of their energy is used to create light. This explains why many companies and consumers are switching to the more energy-efficient compact fluorescent lights (CFLs).

How Do Organisms Obtain Energy?

Adenosine triphosphate (**ATP**), a high-energy molecule, is often considered the "energy currency" of the cell. In other words, ATP is the molecule that cells use as an energy source for performing different types of work such as transporting molecules and ions, building large molecules (that store potential energy), or breaking down large molecules (to release energy). Organisms vary in the ways they create and use energy, but ATP is universal to all of them.

Autotrophs

Plants use sunlight to create energy in a process called **photosynthesis**. Photosynthetic organisms are **producers** or **autotrophs**, because they can create their own food for energy. They are also able to synthesize chemicals and molecules needed for growth, reproduction, and overall maintenance. These chemicals and molecules include hormones, amino acids, fats, and carbohydrates.

Photosynthetic plants contain **chloroplasts**, subcellular structures instrumental to photosynthesis. Chloroplasts contain **chlorophyll**, a pigment molecule that is able to absorb sunlight and use it to create energy. This reaction "builds up" or absorbs energy so we call it **endergonic**—meaning it has a net positive change in free energy (energy available to the organism). Within chloroplasts are enzymes called **rubisco** that "fix" carbon atoms—by transforming them from their gaseous CO_2 form into solid form. The carbon is used to build sugar. Sugar, as we discussed earlier, has potential energy stored in its covalent bonds.

ATP provides much of the energy to drive photosynthesis. The products of photosynthesis are ATP, sugar, and oxygen. Have you ever noticed that it is easier to breathe in a room containing plants? It is because they release oxygen!

In addition to chloroplasts, plants also have mitochondria, a subcellular structure that enables them to break sugar down and use the energy by-product of the reaction.

Chloroplast Anatomy

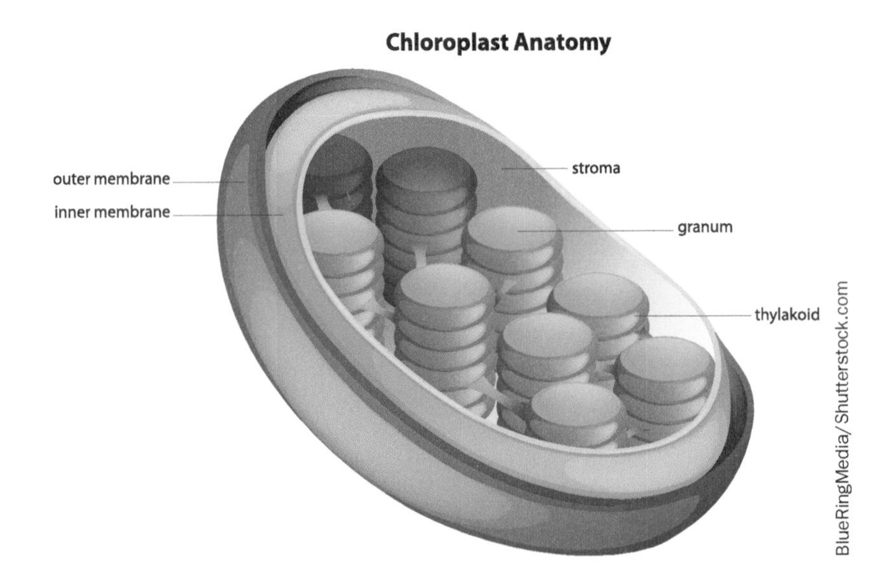

Heterotrophs

Heterotrophs (nonphotosynthetic organisms) are also called consumers because they must consume other organisms for food and nutrients. They also must consume other organisms for many and molecules needed for growth, reproduction, and overall maintenance. These chemicals and molecules include hormones, amino acids, fats, and carbohydrates. They use respiration to break down food and obtain energy. Respiration is a process that *breaks down* sugar molecules (by splitting carbon) for energy. ATP is created in the process. During respiration, there is a net *negative* change in free energy since energy is lost (no more potential energy stored in the carbohydrate bonds), so the reaction is considered **exergonic**. (Please note that although plants are autotrophs, they also respire when they break down sugar for energy).

Aerobic respiration requires the presence of oxygen and takes place in a subcellular structure called the **mitochondria**. Multicellular organisms mostly use aerobic respiration because it generates enough ATPs to meet the energy needs of a complex organism. When oxygen is present, the amount of ATPs generated is ample, so often mitochondria are considered the "powerhouse" of the cell. The cells in your muscles and liver (the liver is the most metabolically active organ) contain many mitochondria because they have high energy needs.

Anaerobic respiration is the process of breaking down sugar in the *absence* of oxygen. Mitochondria are not affiliated with this process. Simple organisms like bacteria and yeast often respire anaerobically. The amount of ATPs generated is modest, but enough to meet the energy needs of simple cells or temporary needs of more complex organisms.

In certain circumstances, our bodies respire anaerobically (in addition to aerobically). Anaerobic respiration in humans typically involves sprinting or weight lifting; short, intense exercise that requires strong bursts of energy. Breathing is often short and quick, and so oxygen has not yet had time to reach the tissue. The ATP-energy payoff in this case is minimal.

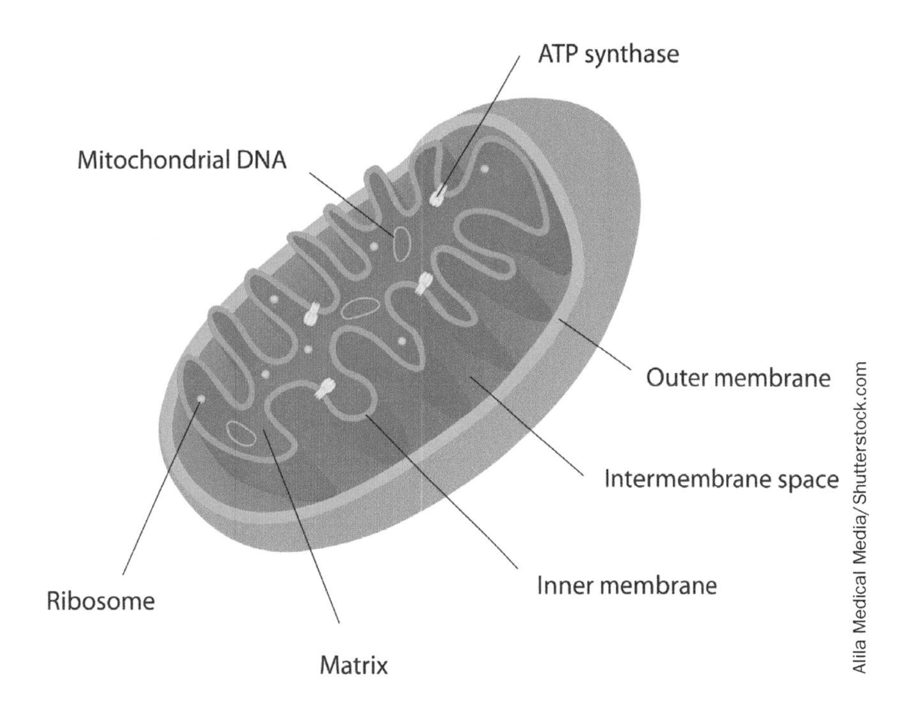

When our "second wind" kicks in, it means we are breathing evenly and oxygen has had time to reach our tissue, and we are now respiring aerobically (in the presence of oxygen).

When we stretch, we not only loosen muscle and avoid risk of injury but also help release some of the waste products of muscle metabolism. This activity is not unique to humans; animals perform this self-medicating strategy as well. Many of our yoga poses (downward facing dog, for one) mimic the postures of big cats and wolves.

Fermentation is a type of anaerobic reaction that creates alcohol as a by-product. Wine is made through fermentation (yeasts ferment on the sugar present in grapes.)

Fun Fact

In earth's early stages, there was very little free oxygen gas in the atmosphere, so anaerobic respiration was a favored means of obtaining energy. Cyanobacteria are small, single-celled organisms that perform photosynthesis and release copious amounts of oxygen gas as a by-product. Many scientists and historians credit cyanobacteria with providing the atmosphere with the high concentrations of oxygen gas in the atmosphere today and therefore facilitating the many different life forms out there that breathe oxygen.[ii]

Review Questions

1. Define an isotope and how they are useful in modern society today.

2. What is the difference between a cation and an anion?

3. Explain the unique properties of water. How does its chemical make-up explain these unique functions?

4. List life from small to large.

5. What is the relationship between Hydrogens and pH?

6. What is an autotroph? What organelle is found in most autotrophs?

7. What is aerobic respiration? What organelle is associated with aerobic respiration?

8. What are the unique properties of life? Why are these properties considered emergent?

9. Discuss the difference between organic and inorganic molecules. Give examples of each.

10. How do water's unique properties explain why it is a major component of the human body? If water had different properties, what might happen to living things?

11. How does the second law of thermodynamics contribute to entropy (chaos) in the universe?

12. Define a covalent bond.

13. Provide several examples of trace elements.

14. Name the 4 classes of biological molecules. What are their functions and where in the body are they found?

Discussion Questions for Class

A. Explain how eating a large pasta dinner the night before a big-track meet or soccer game is providing potential energy.

B. Up to 90% of energy is lost with traditional incandescent light bulbs. CFLs last longer than traditional light bulbs, use less energy, and emit more light. Ecofriendly bulbs are not only more efficient but they help reduce lead, mercury, and cadmium. From a biological perspective, why are these materials hazardous to us and to the environment?

C. How does using fossil fuels as an energy source contribute to acid rain? How can a low pH and a change in acidity affect soil, lakes, streams, plants, and trees?

D. When learning about Environmental Science, why is it important to know about atoms and molecules?

E. Why is homeostasis necessary for life?

F. Why "respiration" is considered an exergonic reaction? Why "photosynthesis" is considered an endergonic reaction? Where in the body would you find a lot of mitochondria?

Your Personal Sustainability Journal

Find a phone app or website that "rates" personal skincare products based upon how natural and ecofriendly they are. An example of an app is "skindeep."

Read the labels of your everyday products such as toothpaste and moisturizer. What percent of their ingredients are organic or natural? Buy products with natural ingredients.

Turn off and unplug appliances when they are not in use.

Add your own suggestions here.

Endnotes

i. Wild Health: How Animals Keep Themselves Well and What We Can Learn From Them. Cindy Engel, Weidenfeld, and Nicolson, 2002.
ii. What on Earth Evolved? 100 Species That Changed the World. Christopher Lloyd, 2009.

Chapter 3

Biodiversity and Evolution

DID YOU KNOW?

Consider that an estimated 71% of the earth is covered by water, with its greatest volume found within the far-reaching oceans. The oceans serve as the largest storehouse for carbon—we call any such storehouse a *sink*. In this case, the oceans act as **carbon sinks**.

When carbon dioxide reacts with ocean water, carbonic acid is formed. As large amounts of carbon molecules are released into the atmosphere as a by-product of burning fossil fuels, high concentrations of carbonic acid accumulate in the oceans. Carbonic acid lowers the pH of oceans causing **acidification**.

To counteract this, oceans use calcium as a buffer to neutralize pH. Calcium is taken from calcium carbonate, which is found in shells on the ocean floor. This threatens any aquatic life with hard shells (such as corals that create the coral reefs). In fact, acidification threatens the livelihood of any aquatic species with a narrow range of tolerance to dramatic changes in pH.

Biodiversity

It is estimated that 99% of the species that once roamed earth have now gone extinct. Currently, estimates of the extinction rate range from 1,000 to 10,000 times its normal rate as a result of habitat loss, climate change, pollution, deforestation, and other human-influenced activities.

Extinction results in a dramatic loss of biodiversity. It is difficult to obtain precise data, but it is estimated that a parcel of rainforest land the size of a football field is cut down every few seconds. The nonprofit Amazon Watch claims that 20% of the Amazon has been cut down or destroyed. It is cut down for grazing land, soy, oil exploration, dams for hydroelectricity, energy crops, to name some reasons for human intervention in the rainforests. In the last

40 years, we have lost 20% of the rainforests, which contain half of the world's biodiversity! Is our addiction to overconsumption worth such a dramatic loss of biodiversity?

In this chapter, we will explore what factors *influence* or correlate with high ecosystem biodiversity. We will also examine compelling incentives to *protect* biodiversity, as well as major existing *threats* to biodiversity.

Biodiversity refers to the various species living and interacting with each other in a given area. This encompasses all individual organisms, as well as their populations and species. It includes genetic diversity and the complex community and ecosystem structures.

Species diversity is the number of different species (see **species richness**) as well as the differences among and within those species. These include all different kinds of species: microorganisms, fungi, plants, and animals. **Genetic diversity** refers to the variety of genes within a species. Genetic diversity is beneficial for a species because it helps the species adapt to a changing environment. Populations are constantly catching up to an ever-fluctuating environment. The more genetic diversity a population contains, the more chances there are for advantageous adaptations to a changing and unpredictable environment. **Ecosystem diversity** encompasses different biological communities, ecological processes, and various habitats. The more ecosystem diversity there is, the healthier and more stable an ecosystem it.

Species diversity can be high, as is the case with the tropical rainforests that contain 50% of the world's species. It is estimated that the rainforests contain over 5 million species of plants and animals! High biodiversity is a sign of ecosystem health and a good indicator that it will recover in a reasonable amount of time after a disturbance or periods of stress.

Areas of high biodiversity are called **hot spots**. The tropical rainforests, coral reefs, and the state of California are considered hot spots. Although not a tropical climate, California is a hot spot because it harbors biodiversity not found anywhere else in the world: Sequoias, Redwoods, Joshua Trees, coastal sage scrub, chaparral, etc.

grmarc/Shutterstock.com

Rainforests have high species richness

Biodiversity can also be low as is the case with the Gulf of Mexico. Here, water quality is so poor and concentrations of oxygen are so low that only simple organisms like **microbes** can grow. Areas with low biodiversity are called **dead zones**. (Please note that in the case of the Gulf of Mexico, poor water quality can be attributed to pollution caused by human activity.)

Biodiversity is further defined by **species richness** and **species abundance**. Species richness refers to the total number of different species present in a given area, while species abundance refers to the number of individuals of a given species that is present in that area.

An area may have high species richness but low species abundance. Take the tropical rainforests again as an example; it has *high* species richness (high species number) but *low* species abundance (a relatively low number of individuals *within* a species). Conversely, other areas such as the Arctic have *low* species richness but *high* species abundance. Relatively fewer types of plant and animal species are found in the arctic, but of those species represented (reindeer, caribou, polar bears, and the arctic fox), there is an abundance of individuals.

outdoorsman/Shutterstock.com

The Arctic Fox. Relative to the rainforests, the arctic has higher species abundance.

Wolfgang Kruck/Shutterstock.com

Caribou

When we think of biodiversity, we tend to think of beautiful and charismatic and colorful **flora** and **fauna**, but we must not forget less charismatic species living in the soil. Earthworms and mushrooms, although not particularly aesthetically pleasing, are just as important to maintaining an ecosystem's biodiversity and health as are the majestic rainforest trees. These species break down decaying matter and are essential for maintaining biodiversity as they enrich the soil, enabling other species to grow and thrive. In fact, some historians even link the scarcity of earthworms to the collapse of the empire! Their irrigation system (they started using saltwater) killed off their earthworm population. A paucity of earthworms meant poor soil quality and therefore few crops. This, in turn, probably weakened their health and social structures, rendering them vulnerable to invasion!

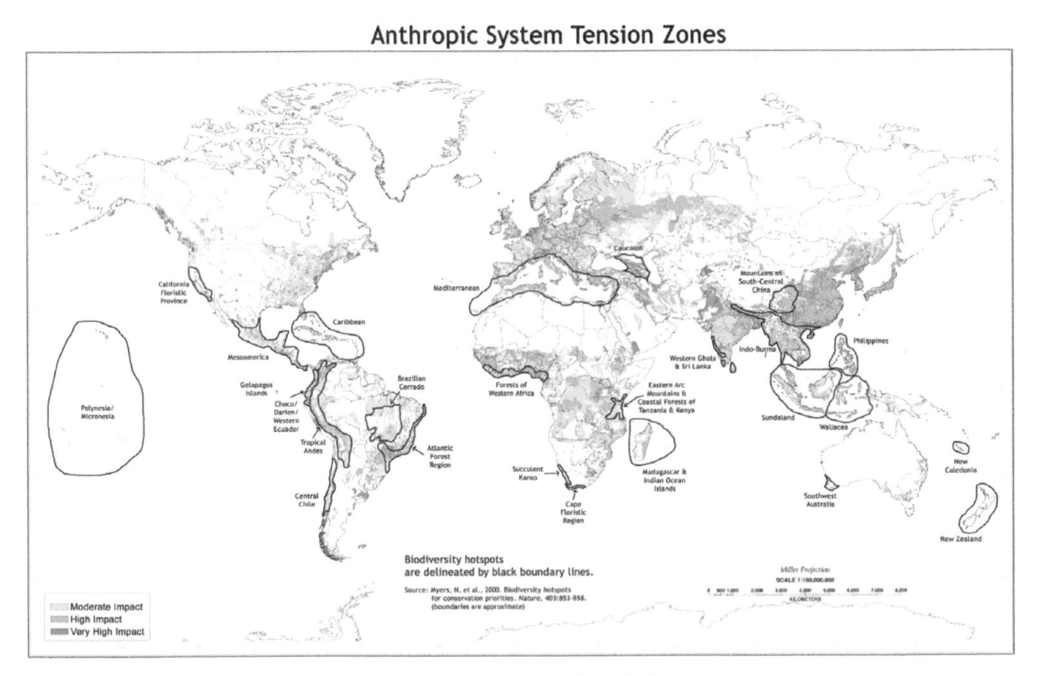

Global Hotspots Outlined in Black. Human impact to Global Hotspots

Source: Land Quality map, USDA-NRCS, Soil Survey Division, World Soil Resources, Washington D.C. Population density map, Tobler, W., V. Deichmann, J. Gottsegen, and K. Maloy. 1995. The global demography project. Technical Report TR-95-6. National Center for Geographic Information analysis. Univ. Santa Barbara, CA, 75pp. Biodiversity hotspots, Myers, N. et al., 2000. Biodiversity hotspots for conservation priorities. *Nature*, 403:853–858.

Factors that Influence Biodiversity

Many factors influence ecosystem biodiversity, but in the following paragraphs we will focus on the most salient variables such as amount of sunlight, age, and size. Let us examine these variables in more detail:

A. The **ambient light hypothesis** states that the *more sunlight* accessible to an ecosystem, the more functional it becomes. More sunlight means more thermal energy, which leads to more reactions and ecological processes. Functional in this context means having many varied and specific roles within an ecosystem (decomposition, photosynthesis, predation, pollination, etc.). More photosynthesis creates more oxygen for an ecosystem. Oxygen is a nutrient necessary for life, so more life is created and thrives. This leads to high ecosystem richness. The more functional an ecosystem, the more biodiversity it harbors, and the more Carbon, Nitrogen, and biomass it creates. A highly functional ecosystem has many plants, herbivores, predators, prey, and decomposers—all forming an intricate and robust food web.

Kim Briers/Shutterstock.com

Consistent with the ambient light hypothesis, high biodiversity is observed near the equator. Because the earth spins on its axis, the equator receives the most sunlight throughout the year. Ample sunlight means more photosynthesis, which means more plants and trees. More plants and trees mean more herbivores, more species that feed on herbivores, and more species that help decompose dead plant and animal matter.

The coral reefs, also found along the equator and also considered hot spots, are sometimes referred to as the "tropical rainforests of the ocean." Coral reefs get an abundance of sunlight and, therefore, are also a highly functional ecosystem.

B. Tropical rainforests, relative to other ecosystems on the earth, are extremely *old*. This factor may also help explain its high biodiversity. One hundred thousand years ago, during an era we call the **Ice Age**, North America was covered thick glaciers. The glaciers eventually receded, leaving lakes, streams, hills, valleys, and new ecosystems in their wake. The glaciers never reached as far south as the equator, leaving the rainforests undisturbed and intact. As a result, the rainforests have a longer historical legacy for the many species that live and flourish today. The millions of different species found throughout this biome have had time to evolve and adapt, undisturbed by glaciers.

Dirk Ercken/Shutterstock.com

The older an ecosystem is, the more death and decaying matter is present. Ironically, more death in this case means more life, since decomposers such as bacteria, earthworms, and fungi break down dead and rotting organic matter, enriching the soil and making it ripe for new vegetation to grow to be consumed by a growing animal population thriving on the abundance. An older ecosystem also means that there is often a significant underground fungal network that facilitates communication among the trees.

C. Hot spots such as the tropical rainforests are often massive in size. *Large land mass* as well as *large amounts of biomass* provide more habitats for species to live and adapt. It also provides oxygen which is a nutrient that enables living things to grow and thrive. They also provide more places to hide from predators or competitors. The result is a lower extinction rate. The tropical rainforest trees grow extremely tall (some up to 80 meters high!), providing niches in which other species can live. In the canopies one finds **epiphytes** (plants that live on the tops of trees), along with the many different rainforest tree frogs and insects. Among the understory live many different snakes, birds, lizards, and even large cats such as jaguars!

Rechitan Sorin/Shutterstock.com

Jaguar

D. There are some places on earth where evolution seems to 'speed up' and create an explosion of biodiversity. In such areas, species venture into new niches and become uniquely and specifically adapted to each niche. An example of this is the finches that Charles Darwin examined while he was exploring the Galapagos Islands (see the paragraph on evolution below). He observed various finches uniquely adapted to eating seeds, fruit, insects, and plants. These finches, he presumed, were so similar that a recent a common ancestor seemed likely. However, they were so uniquely adapted to their environment that he could also not discount that they were distinct species. Such a phenomenon is called **adaptive radiation**. Adaptive radiation often occurs on volcanic islands and mountain tops, where there are ecological islands of biodiversity and species inhabit and adapt to the available unique niche. (Such ecological islands have little gene flow with other populations and so are often threatened by habitat destruction or disturbance. Furthermore, volcanic islands typically have rich soil that serves as sustenance for life.) Criteria for adaptive radiation include plenty of food and resources,

Rangzen/Shutterstock.com

A Black Darwin Finch Atop a Prickly Pear

Kjersti Joergensen/Shutterstock.com

A Galapagos Finch. Note the different morphology from the above finch little competition, and little predation. In other words, there is not a lot of **selective pressure** working the species being discussed (see below paragraph on natural selection).

E. Ecosystems with **intermediate amounts of disturbance** tend to have high biodiversity. (A disturbance in this sense is an environmental factor that compromises the integrity of the ecosystem: storm, fire, flood, etc.) An ecosystem with *too much* disturbance makes it difficult, if not impossible, for species to thrive and adapt, as they are consistently out of their range of tolerance. Too little disturbance, on the other hand, often lends itself to having a single-dominant species that happens to be a superior competitor, to the detriment of other species and biodiversity. An intermediate amount of disturbance prevents overpopulation.

Why Protect Biodiversity?

There is no simple answer to this question. The many factors involved in this discussion topic are, of course, a matter of opinion and subject to debate. Nonetheless, possible compelling responses are both philosophical and practical in nature.

Let us begin with practical considerations: One cannot deny the *economic value* of biodiversity. According to a 1997 study by Robert Costanza, if we were to take on the daunting task of assigning a monetary value to the goods and services that biodiversity supplies humans, it would run somewhere in the ballpark of $44 trillion![i] This value may seem like a staggering amount, but consider that biodiversity provides us with food, crops, safaris, and other materials and services to be sold and exported. Most economies—even energy economies—are built on natural resources from biodiversity.

Biodiversity provides us with wood for our homes and paper to write on. It provides us with recreation, clean water, clean air, clothing, shoes, jobs, perfume, emollients, aesthetics, and medicine. In fact, some species contain chemicals that are anti-inflammatory, antiviral, and antimicrobial. And some have chemicals that act as painkillers!

Biodiversity also provides *ecological value*. It provides us with protection from storms and floods. It moderates climate, provides wildlife habitats, acts as a carbon storehouse, creates food-webs, recycles waste, and maintains an overall balance to the biosphere.

It provides us with a means for eco-tourism in which we travel to exotic destinations and educate ourselves on endemic flora, fauna, and ecological systems. This sometimes stretches our comfort level but also broadens our horizons with its immeasurable educational value. It enables us to feel connected to the earth and landscape and its native people. It also provides us with an ecological sanctuary—somewhere to escape the stresses and burdens of everyday life. The earth's remote pockets of nature are places to clear our heads, breathe fresh air, commune with nature, and gain perspective on our problems and obstacles.

Biodiversity provides us with *scientific value*. It offers areas in which scientists can go, undisturbed by urban noise and distraction, and conduct field research. This inevitably provides us with a wealth of knowledge and inspiration relevant to all aspects of society.

On a more philosophical level, all life on earth possesses its own *intrinsic value*—an innate integrity and unique place in the biotic world. But it appears as though our approach (stemming from our materials economy based on extraction) has morphed into treating biodiversity as though it is fungible—easily replaced or substituted over time. Not only is this approach unrealistic and unsustainable, but it devalues biodiversity and denies its intrinsic value.

Furthermore, there are *cultural* reasons to protect biodiversity. There is a positive link between biodiversity and cultural and societal diversity, as well as the diversity of languages. If we lose our biodiversity, we lose the others linked to it, and cultural knowledge associated with it. So making efforts to preserve biodiversity and live more sustainably ultimately protects many other cultures, their native languages, and knowledge surrounding biodiversity.

And finally, our plants, animals, fungi, microbes—*all* of life's biodiversity—comprise a legacy that should get passed down to future generations so that they too can experience its natural beauty and immeasurable value. We must act as responsible stewards of the environment in order to achieve this.

Threats to Biodiversity

Biodiversity is the result of billions of years of evolution. Now through human activity and intervention, the process is increasing and we are faced with many threats to which would lead to a loss of biodiversity. Some are more obvious and on a grander scale than others. For simplicity, we will summarize the major threats to biodiversity and collectively give them the acronym HIPPCO (using the first letter of each heading to create this mnemonic):

Habitat Loss

Habitat loss is the number one cause for extinction. Our growing human population needs homes, schools, paved roads, and other resources. This means construction, logging, mining, constructing dams, and paving over land inhabited by other species. This ultimately leads to fragmenting their land. Species that do not die out (as a result of habitat and food loss) are driven to other habitats where they encounter new parasites, new competition, and new predators. Such prolonged stress would attenuate the already struggling number of species. The conversion of land for agriculture and development creates an overall loss of biodiversity.

Another cause for habitat loss is unsustainable agriculture. An example of this practice includes cutting down the rainforests to grow African Palm that feeds the growing demand for palm oil. It also includes cutting down rainforests to provide rangeland and soy for cows to supply the world's substantial meat consumption.

Climate change and other major environmental shifts (in pH, salinity, oxygen concentration, water temperature, etc.) can also directly or indirectly lead to habitat loss. Because communities are so interconnected, even a minor shift in environmental conditions can have dire consequences.

Invasive Species

Invasive (sometimes called exotic) species are species from other parts of the world introduced to a new habitat. These transplanted species now lack their natural predators, competitors, and other population controls. Some exotic species prey on or outcompete native species which severely compromises biodiversity. So it should not come as a surprise that the introduction of invasive species is a major cause for the extinction of native species.

Take for example the African honeybee (*Apis mellifera*) which was introduced to Latin America in order to assist pollination. After being transplanted, it soon became evident that this aggressive species of bee often swarms, and its sting is sometimes lethal to humans. Due to its aggressive nature, African honeybees have outcompeted and killed off much of the honeybee population native to South America. Black rats introduced to the Galapagos Islands in the 1800s prey on giant tortoise hatchlings. They have, in part, been responsible for the giant

Fesus Robert/Shutterstock.com

African Honeybees

tortoises having been close to extinction in the 1970s. (With rigorous conservation efforts, the giant tortoises are now stable and breeding in the wild.)

Another example of an invasive species is the lamprey, an eel-like sucker fish accidently introduced to the American Great Lakes. Lampreys prey on native lake trout and have been such successful predators that presently, very few trout can be found in the Great Lakes.

Invasive species can also contaminate the gene pool of native species, ultimately negatively impacting native populations. It is estimated that the economic loss as a result of invasive species is roughly 1.4 trillion in U.S. currency.[ii]

Andrei Nekrassov/Shutterstock.com

Lamprey

Population

All other threats to biodiversity can be traced back to overpopulation. Having over 7 billion people on earth means our resources are exhausted—from land to minerals to freshwater, the earth simply does not have enough bounty to effectively distribute its goods to such a large and growing population.

Humans are not the only ones feeling the pinch from diminishing resources; other species rely on them as well and suffer as a result. Because of the increase in international trade of plants, animals, and food, pathogens are traveling much more rapidly than ever. This means that wildlife, once shielded from pathogens across the globe, are now exposed to new threats at a growing rate. Often they do not have developed immunity to fight off the newly introduced pathogens. Diseases carried by domestic animals and livestock can harm wildlife populations. African wild dogs for example, have been exposed to rabies and distemper. See our chapter on population growth.

Pollution

There are many types of pollution: air, water, thermal, noise, soil, light, to name a few.

Chemicals and toxic waste from our labs, factories, and urban areas leach into our soil and our water supply, and accumulate in the atmosphere.

Wealthy nations have the resources and technology to purify water and air, but what about those living in developing nations that do not have access to clean water or air? Almost 1 billion people worldwide are without clean water and that number is expected to increase.

We must also consider plants and wildlife that also share our land, air, and water and are also negatively impacted by pollution. Species that communicate acoustically, like whales and dolphins, are vulnerable to noise pollution; while other aquatic animals are vulnerable to changes in water temperatures. Pollution is linked to **eutrophication** of bodies of water, causing algal blooms and often leading to loss of loss of fisheries and in extreme cases, dead zones.

Since the Second World War, wild animals have been increasingly exposed to artificial toxins and chemicals. They have no evolutionary history with these toxins, which means there has been little selective pressure on detection. This in turn means that wildlife have little or no evolutionary adaptation to detect and avoid them. Birds readily eat artificial pesticides and have become poisoned. Marine mammals eat fish contaminated with dioxins and DDT. There are many more examples of species vulnerable to pollution, so please see our chapter on Pollution. more information on pollution.

Some of our common apparatus, material and gear that are characteristic and common to human lifestyle can become injurious to wildlife. Barbed-wire fence, derelict fishing nets, and electric cables can all be harmful or even deadly to wild animals. Speaking from an evolutionary standpoint, the wildlife have had little time to adapt to their presence and learn avoidance of harm or injury.

Climate Change

As we discussed in Chapter 1, according to NASA, earth's lower atmosphere has increased 1 degree Celsius since 1880. This has devastating effects on biodiversity. Flowers are blooming before their pollinators can successfully access them. Species are forced to migrate to the poles to find more amenable temperature ranges. Species' ranges are both constricting and contracting. We are seeing greater extinction rates and changes in species composition and interaction. Pathogens are attacking new species that lack the immunity to fight back against them. We are experiencing droughts and other temperature extremes—photosynthesis rates decrease with extreme temperatures, possibly causing ecosystem collapse. We are experiencing a spike in vector-borne (spread by mosquitoes or ticks) diseases that harm not only us but wildlife as well. The melting glaciers and ice caps leave polar bears, which perch atop arctic glaciers while fishing, weakened from exhaustion and lack of food. Melting glaciers also expose carcasses with pathogens to which present day species are not immune.

Overexploitation

Our economy and many others are built on **extraction**—taking vital goods and resources from the earth. This leads to the excess use of vital goods and resources taken from the earth faster than they are able to replenish naturally. Over-fishing, over-hunting and over-collecting of species can quickly result in a loss of biodiversity.

The assembly of an ecosystem and all its inhabitants are vital for the evolutionary process. Because of the complexity of ecosystems, most wildlife and plants suffer from losing such resources. They also suffer from being exposed to the pollution caused by the extraction. It drives them from their homes, forces more competition for resources, and depletes food supply.

Biodiversity and habitat loss can cause some species to be considered formally endangered. **Endangered species** are species with such low abundance that they are at high risk of becoming extinct.

MVP (minimal viable population) is the minimum number of individuals that must be present in a species population for it to be considered at low extinction risk. For most land animals, this number is at least 500. Approximately 20,000 members of a population must be present for biological evolution to occur.

A **threatened species** has a good chance of becoming endangered in the near future. A **sensitive species** is a species that is vulnerable to habitat loss; therefore, conserving habitats should be a top priority in order to preserve the species. We will explore more on endangered and threatened species in later in this book.

Protecting Biodiversity

Drafted in 1992 and enforced in December of 1993, the United Nations created an international treaty called The Convention on Biological Diversity, often referred to by its acronym CBD. Its mission is:

1. The conservation of biological diversity,
2. The sustainable use of the components of biological diversity, and
3. The fair and equitable sharing of the benefits arising out of the uses of genetic resources.[iii]

In other words, the CBD is committed to ensuring that biodiversity and natural resources are used sustainably to avoid long-term biodiversity loss or depletion. It also secures fair access to and benefit from biodiversity and natural resources for *everybody*.

This treaty is seminal because it protects citizens of countries with rich natural resources (often living in poverty) from getting short changed by big companies (from developed nations) that cut down forests and remove natural resources for monetary gain. It offers financial assistance and resources to the least-developed countries (LDCs) and advises on implementation. The CBD would theoretically help alleviate worldwide poverty and help stabilize the economies of countries with rich natural resources.

Indeed, its efforts have been met with marked success: more countries are ratifying this treaty; already there is compelling evidence that deforestation rates in tropical areas have declined as a result of efforts to decelerate it; countries in the Caribbean are increasing their marine and coastal conservation; the Maldives pledged to make the country a Biosphere reserve; and finally, measures to control alien invasive species have resulted in many species being moved to a lower extinction risk category. Because of the uptick in poaching, illegal animal trafficking and overall habitat loss (an public awareness of this), there are many non-profit organizations and other agencies dedicated to preserving biodiversity, rehabilitating ecosystems, conserve species, and fight the illegal wildlife trafficking.

Evolution

What we observe as earth's impressive biodiversity is, in fact, the result of millions of years of evolution. **Evolution** is defined as *the change in gene frequency of a given population over time.* It is important to note that populations evolve, not the individuals within a population.

In 1831 Charles Darwin, an English naturalist, embarked on the British ship *H.M.S. Beagle* to study the biodiversity of the Galapagos Islands of the South Pacific. He noticed that the tortoises were much larger than they were on the mainland. He also thoroughly examined the Galapagos finches for their unusual morphological diversity.

Darwin observed that each distinct species of finch has its own unique adaptations, making it undeniably separate and different from other birds on the island. At the same time, however, morphologically they appeared very similar, and so one could not deny a common ancestor. He could not answer the question why, if species are created for the environment (as previously believed), were there 13 species of birds for one single environment? This observation was the catalyst for further examination of the question: perhaps species *can* change, but how?

In 1859 Darwin published his book *The Origin of Species* in which he describes the evolution of species through natural selection. According to Darwin, the theory of natural selection is based on several observations:

1. There is observed variation in every species population,
2. Within the population, some members of the species will have beneficial variations that will ensure better likelihood of survival,
3. The members of a population with the beneficial variations will leave the most offspring, and
4. The beneficial traits are heritable and passed on to future generations.

In colloquial terms, we call this process survival of the fittest.

The Origin of Species also established that all species descended from a common ancestor through a process called speciation.

Speciation is the process through which a new species arises as a result of adaptation to an ever-changing environment.

Allopatric speciation is speciation occurring in the presence of a physical barrier, such as a mountain range or a river, etc. In such a case, gene flow is blocked, and evolutionary agents work on the now-isolated population, resulting in new species.

Sympatric speciation is speciation occurring *without* the presence of a physical barrier. Often the

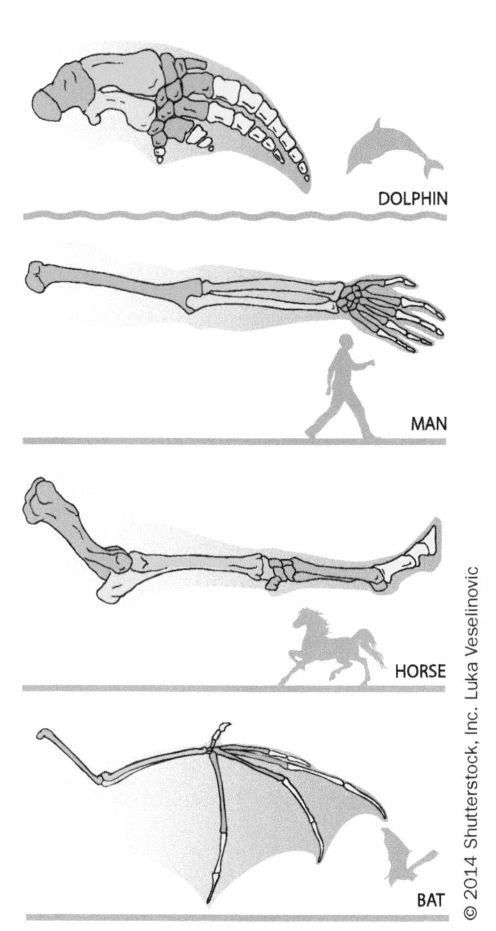

Comparable Arm-bone Orientation of Different Animals, Suggesting a Shared Evolutionary Past.

reason for lack of gene flow between populations is behavioral. In the case of plants, speciation may be the result of self- or cross-pollination, as is the case with plants. Plants have a higher threshold for abnormal chromosome number than humans, and so hybridization is a common occurrence in the Kingdom Plantae. This is an example of sympatric speciation.

Adaptations are structures or behaviors that increase the livelihood or **fitness** of an individual. Over time, structures and genetic frequencies shift so profoundly that a new species forms.

Various anatomical structures and much other observable evidence support Charles Darwin's theory of evolution through descent from a common ancestor:

Narchuk/Shutterstock.com

A. Although sizes and shapes may vary, the arm bones of dogs, cats, whales, and humans all have similar orientations. This suggests they all share a common ancestor with this anatomical trait and evolved separately into new species. Anatomical structures that provide evidence of a shared evolutionary past are called **homologous structures**.
B. Further evidence to support evolution comes in the form of **analogous structures**, similar structures that are the result of **convergent evolution**. Convergent evolution results when unrelated species occupy similar environments and so natural selection favors similar adaptations. This causes the unrelated species to look or behave similarly. These behaviors or appearances are considered analogous. Analogous structures, therefore, are anatomical structures that do not share an evolutionary past but are the result of similar environmental selective pressures. Consider the eyes of a squid and the eyes of humans. They evolved separately as a result of similar selective pressures, although there is no evidence of a recently shared evolutionary past. The fact that both bats and dolphins communicate and navigate through echolocation is also an example of convergent evolution.
C. Goosebumps in humans is left over from a time when we had fur! When the flight or fight response was initiated long ago, the *erector pili* muscles in the skin would contract. This would theoretically make our hair stand on end and so make us look larger to potential predators or other threats. Such a structure is considered **vestigial**; it is no longer employed by the organism but left over from a time when it was. This suggests that humans adapted to their environment, changing genetically and morphologically.
D. The fact that all living organisms, from the bacterium to the cow, share the same molecules of DNA and sequences of amino acids for building proteins is **biochemical evidence** that supports the idea that we evolved from a common ancestor.

E. Sediment such as dead plant and other organic matter settles in layers and eventually forms sedimentary rock. Each layer of sedimentary rock is called a stratum and collectively called strata. Each stratum represents a distinct period of time in earth's geological history. Some fossils are found in one strata of rock but not in another, suggesting that species have thrived during a specific finite period, and then either died out or evolved. This **fossil record**, therefore, gives evidence that species have evolved, adapted, or gone extinct, supporting Darwin's theory of adaptation and descent from a common ancestor.

Michal Ninger/Shutterstock.com

Fossil Trilobite Imprint in the Sediment

F. Finally, there is **Biogeographical Evidence** to support Darwin's ideas of a common ancestor and adaptation. Biogeography is the study of the distribution of living things throughout our planet earth. At one time in earth's history, the continents were connected to make the supercontinent called **Pangaea**. Over time, as a result of **plate tectonics,** the continents gradually shifted and diverged. Species separated from each other and speciation occurred. Species became distinct from each other. Nonetheless, they remained similar in appearance and shared an evolutionary history. An example of this includes the Galapagos finches that remained similar to mainland finches in outward appearance, but evolved into different species. Another example includes the marsupial fossils that have been found in North America, South America, and Australia dating back 30–40 million years ago. There was a land bridge around that time in earth's history connecting South America to Australia, so it is very likely that the marsupials dispersed and evolved into new and distinct species whose fossils we examine today.

Environments are upredictable, are ever-changing, and subject to various fluctuations. Species are constantly evolving to catch up to changing environments. According to Charles Darwin, it is not the strongest species that survives, nor the smartest species that survives. It is the species that is least resistant to change.

Sometimes, having the "highest fitness" is the ability to cooperate with other species or share a resource or home, and avoid direct competition. The oak tree for example (Genus *Quercus*—an angiosperm) is one of the most reproductively successful trees in North America. Each tree is considered the land equivalent of a tropical reef (because of its high biodiversity) and the entire tree is protected in some states. The typical oak is home to over 40 species of insects, 30 species of birds, and at least 200 species of moths. Insects lay their eggs inside the leaves, which in turn provide food and protection. Because these trees have so many visitors and inhabitants, this cooperative nature of Oaks has proven to be quite an effective reproductive strategy. Oak trees are prolific—they can yield up to 100,000 acorns a year and provide an abundance of nutrients to many species, which disperse their seeds. We will examine some more of these examples in Chapter 4.

Review Questions

1. How does the presence of analogous, vestigial, and homologous structures support Darwin's ideas of descent through modification?

2. Compare and contrast "species richness" with "species abundance."

3. What factors make a species vulnerable to extinction?

4. Describe what major factors influence biodiversity.

5. Where might you find a 'hot spot?' How about a "dead zone?"

6. What is the difference between sympatric and allopatric speciation? Please provide examples.

7. Define evolution. What are the four observations upon which natural selection is based?

8. What are major threats to biodiversity?

9. Define 'convergent evolution' and provide an example.

10. What are the major objectives of the CBD?

11. What factors are required for adaptive radiation to occur?

Discussion Questions for Class

A. How have the root causes of environmental problems compromised earth's biodiversity? How may they compromise biodiversity in the next 50 years?

B. We can point out many examples of aggressive and invasive species that have threatened and limited biodiversity: the honeybee population is threatened by Apis mellifera, and lampreys threaten rainbow trout. How might this impact the food chain? Can you think of others?

C. Beyond the $44 trillion economic value of biodiversity are the incalculable economic, medicinal, recreational, and intrinsic values of biodiversity. Do we have a moral obligation to maintain it?

D. Does intrinsic value matter more than economic value? Where should our priorities be as a society regarding these values? Does ecological value matter more than economic value? Why or why not?

E. Is Darwin's Theory of Evolution through natural selection in conflict with the Greeks' idea of provident ecology? Why or why not?

F. Can you think of other examples of homologous structures? Analogous structures? Vestigial structures?

G. How does the presence of analogous, vestigial, and homologous structures support Darwin's ideas of descent through modification?

H. What factors make a species vulnerable to extinction?

I. Describe what major factors influence biodiversity.

J. How can "cooperation" between species increase fitness? Please provide an example.

Your Personal Sustainability Journal

Take a walk around your home, office, or school and take a mental inventory of all wildlife. If there are native bird species, research on the internet to learn about their natural habitat. Now provide an appropriate, small, wildlife home or habitat.

Write down 10 things you noticed (that you used today that came from nature's biodiversity. Now think about how much land and resources were used to provide this service for you. What is involved? Can you limit the use of any of these products?

Cut down on meat consumption, especially red meat. This is the quickest way to reduce your ecological footprint. Try going vegetarian for a week!

Many of the rainforests in Southeast Asia are being cut down for palm oil plantations to meet growing demand in the West. This takes habitats away from Orangutans whose population is threatened, among other species. Educate yourself on food and products that use palm oil. Limit your consumption or buy from companies who have revised their policies regarding palm oil. Some are harvesting palm oil in a way that preserves the rainforests and peatlands.[iv]

Add your own suggestions here.

Endnotes

i. http://www.esd.ornl.gov/benefits_conference/nature_paper.pdf
ii. www.unep.org
iii. www.cbd.int
iv. http://www.ucsusa.org/assets/documents/global_warming/deforestation-free-palm-oil-scorecard.pdf

Chapter 4
Ecosystems and Energy

DID YOU KNOW?

Most energy sources used by humans for heating homes, cooking food, heating water, and powering vehicles—from solar panels to coal to oil to biomass—originate from the sun! Coal is the remains of photosynthetic organisms that once upon a time harnessed the sun's energy for nourishment and energy. Over time, some of their organic remains became subject to stress, pressure, and other extreme conditions, eventually forming coal. All other forms of energy (such as oil which is the product of animal remains) can be traced back to the food web, where it is ultimately fueled and replenished by (you guessed it) the sun!

Ecosystem Energy Flow

According to the **First Law of Thermodynamics**, *energy cannot be created or destroyed; it can only change form.* Millions of miles away, hydrogen and helium fuse and release incomprehensible amounts of energy. We perceive this phenomenon as sunlight. One percent of this energy reaches earth and is used by plants and other photosynthetic organisms for energy. As the First Law indicates, this energy is not destroyed, but rawther flows through our ecosystem in a one-way direction. Energy is the impetus for all ecological functions.

As energy moves from one level of an ecosystem's food chain to another in a one-way direction, it is moving among different energy levels or **trophic levels**. Since they capture the sun's energy directly, photosynthetic plants are among the **first trophic level**. They are the most abundant organisms in the ecosystem, as they harness the highest percentage of the sun's energy. Plants are **producers**, also called **autotrophs**; organisms that make their own energy. They also produce all the chemicals they need for survival, growth, reproduction, repair, and maintenance. These necessary chemicals include amino acids, hormones, fats, and carbohydrates.

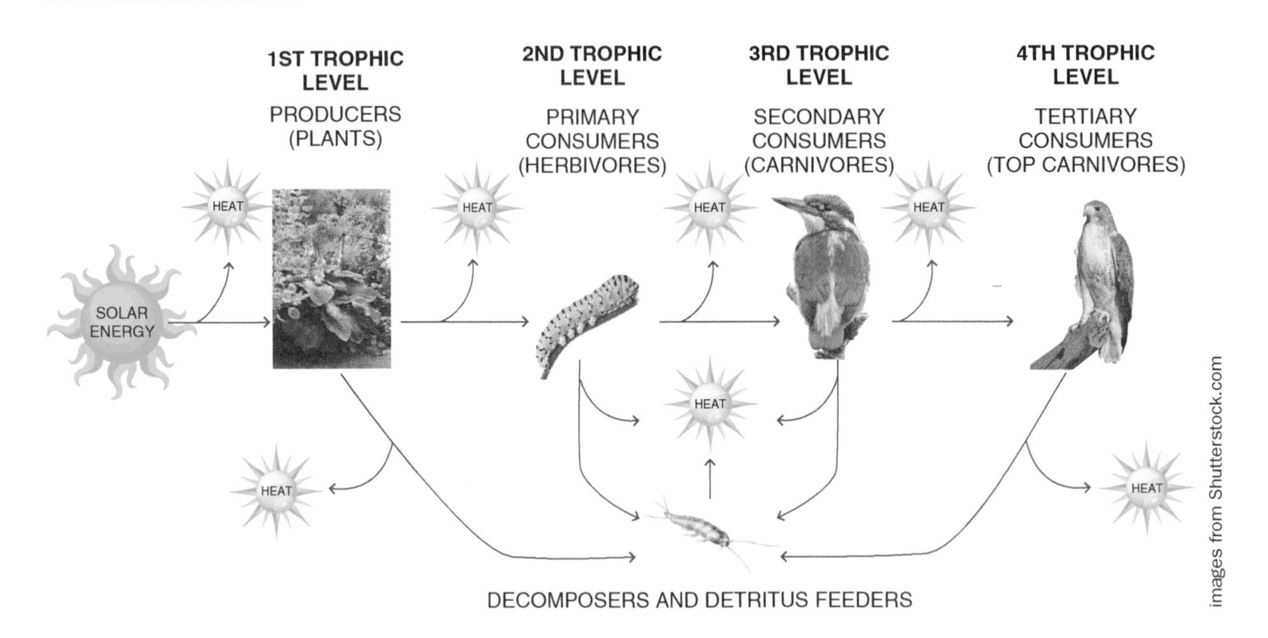

According to the **Second Law of Thermodynamics**, *as energy changes form, most of it is lost as heat*. So as energy moves from one organism to another, not all energy is retained. In fact, 90% of it is lost as heat! This means that herbivores only obtain 10% of the energy stored in the chemical bonds of plants. Herbivores are **primary consumers. Consumers**, also called **heterotrophs**, are organisms that must consume *other* organisms for energy and for all the chemicals needed for survival, reproduction, growth, maintenance, and repair. A deer is an example of both a consumer and a heterotroph because it must eat plants for energy and nourishment. In aquatic ecosystems, algae-eating small fish are both consumers and heterotrophs. We *Homo sapiens* are consumers of things such as starch, grains, nuts, and beans for energy and also to obtain essential fats and amino acids from other food sources. As consumers, we do not produce all the energy, molecules, and chemicals we need for survival, so we must consume them elsewhere.

Carnivores, or **secondary consumers**, then consume herbivores for energy and nourishment, obtaining only 10% of the energy from herbivores. Carnivores that eat other carnivores—such as lions or leopards or hyenas—are **tertiary consumers**. They only obtain a small percent of the sun's original energy! Tertiary consumers are not among the most abundant organisms in the biosphere because they require so much energy from an ecosystem. **Detrivores**, such as beetles and maggots, feed on dead organic matter. The **decomposers**, such as fungi and earthworms, are the recyclers of an ecosystem. They break dead organic matter into simple atoms and release them back into the environment.

Ecosystems can be either **terrestrial** or **aquatic**. A terrestrial ecosystem is found on land and is often characterized by its type of biome. **Biomes**, the highest point of an ecological system, are expanses of land that are defined by their dominant vegetation which is, in turn, is influenced by climate. Vegetation is the dominant life form in biomes. For instance, a grassland biome is characterized by its grass, although there may be herbaceous types of vegetation such as goldenrod or sunflowers, while deciduous forest biomes are characterized by their trees that annually lose their leaves. Vegetation harnesses the highest percentage of the earth's energy relative to other life forms and requires the least amount of energy from an ecosystem. We will discuss biomes in more detail in Chapter 10.

Shutterstock.com

Deer are Herbivores and therefore Primary Consumers

Aquatic ecosystems are either freshwater or marine/saltwater. Adaptations of wildlife found in freshwater ecosystems are different than that of those found in marine ecosystems, since the way freshwater fish maintain water concentrations vary from how saltwater fish maintain theirs. Most marine bony fish have body fluids less concentrated than surrounding seawater. As a result, they lose water through their gills to surrounding seawater. In contrast, freshwater bony fish have body fluids that are more concentrated than the surrounding water medium. Water flows into their gills while salt and other solutes are constantly diffused out through their gills. Nonetheless, both are threatened by overfishing and habitat destruction. It is predicted that a high percentage of marine and freshwater fish will go extinct in the near future.

Avik/Shutterstock.com

Lions are Tertiary Consumers

Chemical Cycling

There are several inorganic molecules that serve as nutrients necessary to sustain life found in abundance and in reservoirs throughout our ecosystem. These molecules, along with energy, flow in a one-way direction. They cycle through soil, plants, animals, protists, fungi, and minerals. All life on earth is connected to these molecules in one form or another. A few of these molecules are discussed below. The nutrients are part of CHNOPS and touch most life in many capacities. As we examine each molecule, please ponder the interconnectedness of life with each of these nutrients. As you read through the cycles, please start thinking about how human activity has taken each of these cycles out of their natural balance, and what we can do minimize our human impact.

H_2O

It is natural that we would start with water, since life is composed mostly of this ubiquitous molecule. Our bodies are made of up to 75% water, and it is the only substance on earth found commonly in a liquid, solid, and gaseous state. Our knee-jerk response is probably not to view water

leonid_tit/Shutterstock.com

as scarce or as a precious commodity, since we see it everywhere and perceive it to be abundant. Earth is covered by more than 1,400 million cubic kilometers of water, averaging over 70% of the earth's surface 97% of the world's water is in the ocean, with over 321 million cubic miles in the ocean. (source NOAA). Because it is so abundant, water is considered a renewable resource. But this perception can be misleading, since less that 1% (source NOAA). of the world's water supply is available to us as drinking water (the rest is either saltwater found in the oceans or frozen in the form of ice glaciers.) We consume it faster than it is purified as freshwater. For this reason, freshwater is considered a nonrenewable resource.

Water changes from a liquid form to a vapor as it evaporates over the oceans. Most water vapor in the atmosphere is traced back to water evaporation over oceans, but also evaporates over terrestrial sources, just less so. In a process called **transpiration**, water exits pores in the leaves of plants and trees and enters the atmosphere from land. Water vapor constitutes just less than 5% of all gases that make up earth's atmosphere. Water eventually enters the ecosystem when it descends as precipitation in the form of rain, snow, etc. After it reaches earth, water percolates down into the ground becoming **groundwater**. Eventually water seeps further down, settling in solid limestone rock cutouts underground called **aquifers**. Aquifers supply us with much of our drinking water and are an important component of our groundwater and overall water cycle.

Water cycles about and purifies itself anywhere from a few days to a thousand years, depending on the water molecule and where it ends up. Atmospheric water takes roughly 9 days to cycle, while the oceans take approximately 3,000 years. Furthermore, a water molecule stuck in an ice glacier may be very slow to recycle! The water cycle is a natural phenomenon and occurred before humans inhabited the planet and would continue without is. But the presence of humans has altered the hydrologic by contributing to the amount of evaporation. Our unsustainable use of freshwater—from flood irrigation of crops, to excessive watering of lawns, to our overuse of water in our homes—has resulted in an acceleration of water's natural cycle (the hydrologic cycle). Our overuse of water and urbanization (the paving over of natural habitats) contributes to excessive runoff, causing much of water to evaporate before it can be stored as groundwater. The more water we use, the less freshwater is available for human consumption. When it evaporates—then precipitates, water eventually ends up in the oceans where it is not available as drinking or freshwater.

We will discuss water in Chapter 9, but it would be beneficial to start thinking about ways to consume less water. There are the typical personal choices we can employ by taking shorter showers and not leaving the water running for unnecessary periods of time. Native gardens and landscapes that use less water are helpful. But changes in water consumption for industrial agriculture or shifting to a plant-based diet would be the most impactful.

Carbon

Carbon's unique tetravalent properties help explain why it is the backbone of life. It is a ubiquitous molecule that is found in the atmosphere, in the earth's crust, in the biosphere as biomass, in the soil, and its largest carbon sink in the oceans. It is often chemically linked to oxygen to form carbon dioxide, and hydrogen to form methane. The carbon cycle includes carbon in its atmospheric gaseous carbon dioxide form as well as its solid form as it is found in most living organisms. Therefore, carbon cycles from its inorganic carbon dioxide form, into its organic biomass form, then back again. Carbon dioxide gas constitutes a meager 0.039% atmospheric carbon dioxide can have unpredictable side effects on the environment and earth's climate. Source: NOAA (Carbon dioxide is a greenhouse gas with high **persistence.**) We are currently experiencing climate change and weather pattern extremes as a result of increased carbon dioxide in the lower atmosphere.

Life, as we discussed in Chapter 2, is comprised of organic molecules that are carbon based. So how does carbon transform from being an atmospheric gas to being the building block of life? Photosynthetic organisms contain the enzyme **rubisco** that has the ability to fix carbon, or convert it into a form that autotrophs utilize to create carbohydrates and other organic molecules. Photosynthesis is an **endergonic** reaction. Plants are building up energy by taking in carbon and assimilating into their plant biomass. This is called **primary production**. Eventually, herbivores eat plants (or algae) thereby obtaining carbon, and assimilate it into their body. This is called **secondary production** and is performed by **heterotrophs.** Carnivores eat the herbivores, driving the carbon cycle up the food chain.

When plant and animal matter die, decomposers break down their dead remains. Carbon dioxide gas is formed as a by-product of this decomposition, and carbon is returned to the atmosphere in its gaseous CO_2 form. Carbon is also returned to the atmosphere as a by-product of cellular respiration (an **exergonic** reaction) in all oxygen-breathing organisms, as methane gas CH_4 (from cow waste), and by the burning of trees, fossil fuels, and biomass (carbon monoxide is also a by-product of forest fires).

The carbon cycle is a natural phenomenon and occurs whether humans are present or not. However, our consumption and lifestyle have changed the carbon cycle dramatically. With the onset of industrialization came the excessive burning of coal, oil, and other fossil fuels. It also brought forth excessive forest-clearing for range lands and raising of cattle (which release methane) for meat production. These activities have accelerated the natural carbon cycling and contributed to increased concentrations of carbon dioxide in the atmosphere and acidosis of our oceans. We will explore climate change and greenhouse gases in Chapter 12.

Controlling and reducing the amount of carbon involves cutting down on carbon energy sources and phasing out fossil fuels like petroleum and coal. It also involves cutting down on methane, a greenhouse gas that is more potent than carbon dioxide. We can reduce the amount of atmospheric methane by reducing our use of natural gas and our dependency on cattle (by consuming less beef and dairy products).

Nitrogen

The entire nitrogen cycle encompasses nitrogen as it exists as an atmospheric gas as well as its solid form in all living organism. Nitrogen is a major component of the biological molecule DNA, our genetic blueprint. It is also part of CHNOPS.

Nitrogen gas comprises just 78% according to NOAA of all atmospheric gas but is not accessible to humans in this form. So how does it enter our ecosystems and into our cells? Many flowering plants called *legumes* have bacteria living on their roots called nitrogen-fixing bacteria that have the ability to convert nitrogen from its gaseous form into nitrogen compounds such as ammonium (NH_4) and nitrate (NO_3). These compounds are in forms the plant can easily absorb and assimilate. Nitrogen-fixing plants, such as alfalfa and pea plants, are beneficial to agriculture and have traditionally been planted alongside other crops for healthy, fertile soil (preindustrial agriculture days).

Herbivores eat the plants, facilitating the entrance of nitrogen into the ecosystem. Eventually when dead organic matter dies, decomposers break down dead organic matter, and nitrogen gas is re-released into the atmosphere in a process called **denitrification**. It is also released back into the atmosphere through burning fossil fuels.

Like the other nutrient cycles, the nitrogen cycle has been significantly altered by human behavior and consumption. Nitrogen is commonly found in natural or chemical fertilizers and sewage. Chemical fertilizers and animal manure are often applied to agriculture to add nutrients for growth. Excess nitrogen can get caught in our runoff and enter our watersheds, eventually being deposited into lakes and streams that cause algal blooms. This thick layer of algae blocks sunlight from entering the aquatic ecosystem, preventing photosynthesis. As algae die, decomposers proliferate to break down the excess of dead algal biomass, thus depleting oxygen from the aquatic environment. This is called **eutrophication**. According to a 2019 study out of the University of Minnesota, greening of lakes or eutrophication will increase atmospheric methane by 30 to 90 percent in the next 100 years, further accelerating the impacts of climate change.[i] This causes fish kills, a phenomenon in which a large number of fish suffocate and die due to lack of oxygen.

Nitrogen can also get into groundwater and contaminate aquifers. Too much nitrogen in drinking water can be harmful to both infants and young livestock. Furthermore, when fossil fuels are burned, Nitrogen is released and creates Nitrogen Oxides, a major contributor to acid rain which is damaging to agriculture and ecosystems.

Reducing nitrogen pollution would theoretically involve updating and expanding wastewater treatment facilities, and implementing stiffer regulations of these facilities so less nitrogen effuses. There should be more control on nitrogen from fertilizers and disposal of animal waste. And finally, a reduction in fossil fuels that release nitrogen oxides and contribute to acid rain would be preferable.

Phosphorus

Phosphorus is one of the only cycling chemical that is only found throughout the biosphere in its solid form. It is not found in an atmospheric gaseous state as the above mentioned molecules are. Phosphorus is a necessary component of our teeth and bones and is also necessary for energy production. Phosphorus is a major component of DNA, ATP (the energy currency of the cell) and cell membranes (as *phospholipids*).

Phosphorus is found naturally in rocks. As rocks weather and wear down (attrition), phosphorus is released into the soil and eventually taken up as inorganic nutrients by plant roots in a process called *cation exchange*. Herbivores eat these plants, causing phosphorus to flow through the ecosystem in a one-way direction. When organic matter dies, phosphorus is returned to the soil. Phosphorus is considered a limited factor in the ecosystem because it is found in small concentrations. It takes rocks a long time to attrition and release phosphorus into the soil.

Human activity and consumption has altered and increased the phosphorus cycle. Phosphorus has traditionally been found in detergents, but is now decreasingly so. It is commonly found in organic waste from sewage, manure, fertilizer, industrial wastes, and other household products. Along with nitrogen, it is indirectly responsible for eutrophication and fish kills.

Like nitrogen, reducing phosphorus pollution might involve updating and expanding wastewater treatment facilities and implementing stiffer regulations for such facilities. There should be more control on phosphorus from fertilizer sources and during the disposal of animal waste.

Sulfur

The complete Sulfur cycle is still being studied and rounded out by scientists, so there are still gaps in our knowledge. Sulfur is a component of CHNOPS and an important element in amino acids. Like Nitrogen and Carbon, it cycles globally through the biosphere—from reservoirs in the soil to a gaseous state in the atmosphere. Its sinks are found on land, freshwater, sedimentary rock, sea water, and the atmosphere. The largest reservoir of Sulfur is in the earth's crust, stored as pyrite or gypsum. Initially, Sulfur is mineralized, as it transforms from organic to inorganic material. It is released into the soil as rocks weather and attrition. It undergoes oxidation to become Sulfate (SO_4^{2-}). Plants, microbes, bacteria, and fungi reduce the sulfide to a sulfhydryl (SH). From here, it is assimilated into plants and microbes.

Sulfur dioxide, as we examined in Chapter 2, is increasingly released into the atmosphere as a result of the coal industry and internal combustion engines. It forms sulfuric acid, which creates acid rain. As we mentioned in previous sections, this is very damaging to forests, soil, wildlife, and agriculture.

Now that we have discussed energy and chemical/nutrient flow throughout an ecosystem, we will be discussing the very species that each of these energy and nutrient components touches and consumes for sustenance. We will also discuss how different species interacts with each other and its unique place in the ecosystem.

Community Interactions

Now we will discuss how populations interact with each other and the many varied role that each of them has in the ecosystem. A **niche** is a species' place in the biotic world and its role in an ecosystem. It encompasses its nutrient and energy requirements. A **fundamental**

niche is the full potential range of all factors that a given species would encounter in its biotic environment during the course of its lifetime. These factors include, but are not limited to the following: altitude, pH, diet, and thermoregulation. A **realized niche** is a species' place in the biotic world and is often much smaller than the fundamental niche. This realized niche is affected by **limiting factors** such as food availability, predators, competitors, and parasites. (**Limiting factors** are variables that limit population size.) Limiting factors are part of the environmental resistance that pushes back against growing populations. A **specialist species** inhabits a narrow niche range and is usually sensitive to environmental conditions and abrupt change. Most tropical species are specialist species and so are made vulnerable by deforestation. A **generalist species** can inhabit a wide niche range, and so can withstand many changes in environmental conditions. Examples include cockroaches and mice.

A **community** is defined as the many different populations of species interacting in a given area. Populations interact with each other in many different and complex ways. This complexity offers a type of stability and balance to an ecosystem. One form of community interaction observed in an ecosystem is **competition**—when species compete for space, mates, food, or other resources. When two species occupy or compete for the same niche, one of two scenarios may transpire:

1. The species that is the superior competitor will drive out or kill off the less fit species in a scenario we call **competitive exclusion.**
2. The competing species may partition their resources, occupying a smaller niche to avoid directly competing for resources. For example, there are several different species of tree warbler; the Yellow-rumped warbler, the Black-throated green warbler, the Blackburnian warbler, the Bay-breasted warbler, and the Cape May warbler which all inhabit evergreen trees in the state of Maine. Animal behavior is often a case of maximizing benefits while minimizing costs. Instead of directly competing within the evergreen, they occupy

Paul Reeves Photography/Shutterstock.com

Yellow-Rumped Warbler Perched on an Evergreen Tree

smaller niches within the trees. In such a scenario, the warblers occupy a smaller realized niche than a larger fundamental niche. In other words, the warblers are minimizing the costs of competing with other species, while maximizing resource benefit.

Another type of community interaction is **symbiosis**. The prefix *sym* comes from Greek origins and means together while bio means living. Therefore, the word symbiosis is defined as two species of organisms living together. In a scientific context, this means that the two species are ecologically linked. **Mutualistic symbiosis** occurs when two species that live together mutually benefit from this arrangement. Mycorrhizae, for example, are a type of fungus that lives on plant roots. Mycorrhizae help plants take in nutrients from the soil more efficiently. New studies also suggest that mycorrhizae help plants communicate with each other to warn against potential herbivores or pests. Meanwhile, the plant provides the fungus with nutrients from its starch reserves. Both the plant and the Mycorrhizae mutually benefit from this symbiosis.

It is common for Wrasse fish swim in and out of the mouths of moray eels in the Indio Pacific coral reefs. The dangerous eel allows the fish to swim in and out of its mouth, collecting and feeding on diseased tissue and ectoparasites. This cleaning ritual is mutually beneficial to both species.

Parasitism is a type of symbiosis in which one species benefits while the other is harmed or sickened. If you have ever experienced strep throat, or any type of infection, you became the victim of a parasite. In the case of strep throat, the parasite is the *Streptococcus* bacterium that exploits its host (in this case, you) in order to live and proliferate. By weakening its host, a parasite can better take advantage of it as a resource. Parasites can hit the human population hard. Dysentery, an intestinal infection caused by a single-celled amoeba parasite, affects 40 million people per year and kills 40,000. It is spread through food and water through fecal contamination.

Wild and domestic animals are infected with many parasites, such as roundworms, nematodes, and heartworm. Domesticated animals are continually given deworming medication. The Kingdom Plantae is not immune to parasites. Often vines, fungi, or insects can be parasitic to trees, plants, and crops.

Commensalism is a form of symbiosis in which one species benefits while the other is neither harmed nor benefited. On occasion, small crustaceous marine organisms called barnacles attach themselves to whales. Whales are so large in comparison to these small creatures that they do not seem affected, while barnacles benefit by having an environment on which to live. Tropical rainforest trees can grow to 80 meters tall. They often have **epiphytes** growing on them. Epiphytes use the host tree for assistance to reach the sun. It does not harm the host tree, but the epiphyte benefits, so this is another example of commensalism.

Cascuta Americana, a Parasitic Climber

Amensalism is a type of relationship in which one species is harmed and the other is unaffected. A classic example is the black walnut tree (*Juglans nigra*) that produces chemical compounds in its roots that alter the soil composition, making it inhospitable to growth of other vegetation. In this example, the black walnut tree was neither harmed nor benefited, as its root system is established, but the growth of a tree or shrub is inhibited. Penicillin happens to weaken or kill bacteria, although it does not benefit from this. This is another example of amensalism.

Another form of interaction is **exploitation**. Exploitation occurs when one species exploits another for its own fitness and benefit. Parasitism, as discussed above, is a form of exploitation. Predation (the act of one species hunting another for prey) is a form of exploitation as well. Herbivory, another type of exploitation, occurs when a primary consumer (deer for example) eats plants for food. In this contextual example, the deer is exploiting the plant. Herbivory is not considered as predation since a plant may be eaten but does not necessarily perish.

Species are not completely defenseless to exploitation. Even plants have means to protect themselves against herbivory. Milkweed and some Euphorbs (shrubby plants) have a milky substance that is toxic to herbivores, dissuading herbivores to use them as a food source. Plants have also been known to secrete chemicals that *attract* the predator (bats or dragonflies for example) of the insects that are eating its leaves. Some plants contain the chemical *saponin* that protect it from parasitic fungi, insects, and bacteria. Some plants contain alkaloids, nicotine, and caffeine—organic compounds that can be toxic to animals. Monarch butterflies, whose larva consume milkweed, are poisonous to birds. Even acorns of the majestic oak are equipped with the chemical tannin that causes digestive irritation in animals if they consume too many! But eaten in the right amount by squirrels, tannins have many beneficial properties—they are antibacterial, antimicrobial, and antidiarrheal. This suggests that the adaptive benefit of tannin is for dispersal.

jo Crebbin/Shutterstock.com

The Barnacles on a Whale—an Example of Commensalism

The beautiful coloration of Monarch butterflies, called **aposematic coloration**, acts as a warning to potential predators that they contain poison and are therefore unpalatable to potential predators. Small colorful frogs found in the wild tend to be poisonous, their bright coloration advertising that they are dangerous to potential predators. The poison dart frog found in South America has brilliant blue coloration, which acts as a warning.

The Viceroy butterfly of North America looks much like the Monarch butterfly. Both are toxic and both receive the benefits of each others' warning coloration. It is protection since theoretically a bird or other predator only has to learn one time—and relatively quickly—that this coloration is unpalatable. This is called **Mullerian Mimicry**.

Lightspring/Shutterstock.com

A Monarch Butterfly

Alfredo Maiquez/Shutterstock.com

A Poison Dart Frog

Sometimes a harmless species, such as the California King snake, has warning coloration but is in fact not poisonous. This nontoxic snake benefits from the aposematic coloring of other poisonous species, although it is not poisonous itself. This is called **Batesian Mimicry**. Some caterpillars—such as the Spicebush Swallowtail, look like snakes!

Dean Evangelista/Shutterstock.com

The Larva of the Spicebush Swallowtail Resembles a Snake

Coevolution occurs when two species put selective pressure on each other and thereby influence the other's course of evolution. Coevolution often occurs between a plant and its pollinator. The nectar quality of the flower may affect the behavior and fitness of the pollinator, while the behavior of the pollinator may affect evolution of the flowering plant by increasing its fitness. For example, flowers pollinated by hummingbirds typically have copious amounts of nectar with just the right amount of sugar to meet the high metabolic needs of a hummingbird. Their petals also tend to be red. In light of coevolution, it should seem to be no coincidence that hummingbirds are most spectrally sensitive to the color red. The adaptations of a hummingbird-pollinated flower ensure successful pollination.

Sari Oneal/Shutterstock.com

A Viceroy Butterfly

Matt Jeppson/Shutterstock.com

A California King Snake

Coevolution can also manifest as an 'arms race' between predator and prey. Wolves tend to hunt hooved animals such as deer. Wolves have adaptations such as speed that enables them to hunt more efficiently. Deer counteract this by evolving behaviors such as running faster, to avoid predation, resulting in an evolutionary arms race.

Another example of two species that coevolve is bats and moths. Bats hunt through echolocation; they emit a high-pitch sound and the corresponding echo indicates the presence of prey. Moths have counteracted this by employing erratic behavior; dropping to the ground and fluttering to avoid detection. Bats have counteracted this by honing their echolocation skills and have evolved to use better detection methods.

A common mechanism to protect oneself against potential predation is **group living**. If you take a moment to consider, you would recognize that many species live in groups; a herd of sheep, gaggle of geese, school of fish, flock of birds, murder of crows, etc. Why do we observe such behavior if natural selection favors individual interests? The easiest way to answer this is to say that the benefits of group living outweigh the costs. Some benefits that ultimately lead to an increase in individual fitness include cooperation in foraging and foraging success, as well as a reduction in predation risk. If you are among many other members of a population, the odds are in your favor that it is not you who gets eaten or attacked. There are also more individuals to spot potential predators. The honey bee population is considered eusocial and engages in a lot of cooperative behavior. Some of the worker bees clean the hive and keep nursery cells clean. They are able to detect diseased larva or pupa and quickly remove it to keep the disease from spreading through the hive.

Another benefit of group living is **kin selection**: the more related a population is, or the more genes they share among them, the more of what humans interpret as 'altruistic' behavior. When a possible predator is detected, ground squirrels for instance will sometimes release a warning call, implicating themselves and placing themselves in possible danger. However, it is observed that ground squirrels will only emit a warning call if it is their close relatives are in harm's way and therefore their genetics are at risk. When they detect a snake, a potential predator, kangaroo rats will sound an alarm by drumming their small feet on the ground as a warning to others.

Group living is not without its costs, however. Some cons of group living include disease, parasites, food scarcity, and even extra-pair copulation (birds are known to be unfaithful to each other!)

Intersexual selection occurs when one sex of the species, typically the female, is choosy and looks for suitable traits in the opposite sex. That is why it is quite common in the animal kingdom for males, such as birds and fish, to be more colorful than females or have features, such as antlers plumage or tusks, that are absent in females. They are displaying impressive traits. The proverbial plumage of the peacock is an example of male display to attract females. The majestic mane of a lion is also a form of display.

Intrasexual selection occurs when members of the same sex compete with one another, usually for mates. An example of intrasexual selection would be male lions competing with other male lions to take over a pride. Because stakes are high (a victory means successful mating and therefore getting genes in the next generation), these competitions are often bloody and lead to the death.

A **keystone species** has an inordinately large effect on the biodiversity of an ecosystem, such that if they were to reduce in population size, the biodiversity of an ecosystem would be greatly compromised. Top predators and pollinators are often keystone species. For example, if the wolf population were to reduce in size, the deer population would increase dramatically causing rampant herbivory and loss of plant life. Wolves hunt deer and maintain a healthy deer population, tempering the amount of herbivory. Wolf presence therefore causes a trophic cascade and brings biodiversity back to an ecosystem. Without plant life an ecosystem could not thrive. This partially explains why, in addition to top predators, most pollinators (bees, hummingbirds, beetles) are considered a keystone species. Elephants are considered a keystone species as well. As the come through the savannah they knock over trees and upturn soil, helping promote healthy bush growth from which all all species benefit.

A **foundation species** creates habitats for other species. Aquatic polyps, sea anemone-type creatures related to jellyfish, secrete sodium bicarbonate as a protective shield against predators. Polyps eventually die but the protective shield remains, which are in fact the corals of the coral reefs. These corals act as homes for all kinds of marine life, such as fish, sea sponges, and crustaceans. Collectively, they become the coral reefs.

Beavers engineer ecosystems by creating dams. Dams provide niches and homes for ducks, reptiles, fish, otters, amphibians, and others. These species interact with each other in their ecosystem which was initially created and engineered by beavers. Beavers are therefore considered a foundation species.

An **indicator species** is extremely important to an ecosystem because it is an indicator of ecosystem health. Amphibians, such as frogs, breathe through their skin and are therefore extremely sensitive to water and air pollution. In the event that a frog population should start to decline, as it has been observed in recent years, warning bells sound that the ecosystem's health is compromised. We have been experiencing a worldwide decline in amphibian populations, a global outbreak in amphibian fungal infections and an increasing number of frogs are being born with abnormalities. This suggests that their ecosystem is being compromised, and in turn they are being harmed, by the usual root causes—ozone depletion, agrochemicals, and climate.

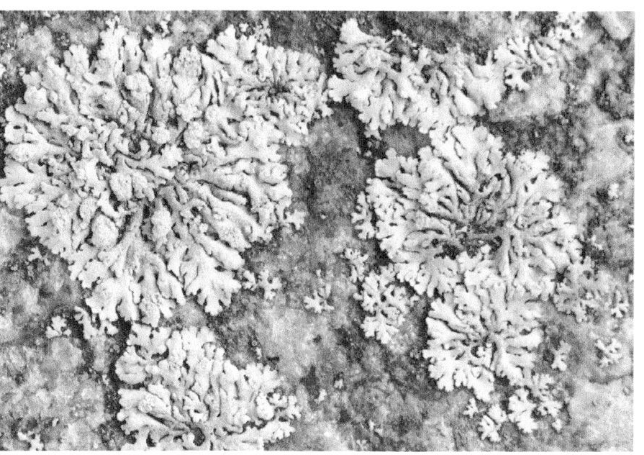

Lichens Indicating Good Ecosystem Health

Salmon, possessing the acute ability to detect small changes in water chemistry, will not swim up river if they detect mining waste such as zinc, or copper. The presence of these elements and the salmon behavior collectively indicate a negative change in ecosystem health.

Lichens—an organism that is the result of a mutualistic symbiotic relationship between fungus and algae—are also an indicator species. In areas with clean air, they will grow and become bluish green. In areas with substantial air pollution, such as urban areas, they will cease being photosynthetic. The fungus will subsequently eat the algae, altering the morphology of the organisms. In some areas you will be able to note this change—often they will grow to become a rust color. Lichen coloration indicates ecosystem health.

Doikanoy/Shutterstock.com

Lichen

Corals expel the algae with which they live symbiotically when they are stressed. This causes the coral reefs to lose their brilliant colors and hues. Coral bleaching, as it is known, is a sign of ecosystem disturbance. Corals are therefore an indicator of ecosystem health.

Umbrella species typically have a broad home ranges and inhabits multiple habitats. They serve as protection for other species, such that when conservation efforts are made to conserve an umbrella species, habitats for other species that share its ecosystem are persevered as well. The northern spotted owl is an example of an umbrella species. When conservation efforts are made for the northern spotted owl, the U.S. old-growth forests are conserved (where this owl species dwells) and therefore all species that inhabit them.

C.M.Corcoran/Shutterstock.com

The Northern Spotted Owl Serves as an Umbrella Species for the Old-Growth Forests of North America

A **Flagship species** is one that appeals to the public's sentimental side. It is a symbol for a particular conservation effort and gives people incentive to raise awareness, participate, and contribute to its continued conservation. Polar bears are an example of a Flagship species because they have become a symbol for the Arctic conservation efforts, the adorable Panda Bear has become a symbol for China's conservation efforts of their bamboo forests, and African Lions and Elephants have become symbols for the dwindling number for megafauna on the African Continent.

Konstantin Novikov/Shutterstock.com

Polyps

Upon conclusion of reading this chapter and how chemical cycling reaches all life in some form or another, and how all species interact in a community, I hope you can appreciate how interconnected a given ecosystem is. If any disturbance, either natural or human-caused, shifts an ecosystem out of balance, all chemicals, plants, animals, and microbes are affected as well. Understanding this balance is the first step in preserving our natural habitats.

Growth Curve Trends

Species fall into two distinct categories when summarizing growth curve trends, R-selected species and the K-selected species. However, most species circle around both growth curves.

R-selected species procreate in litters, are opportunistic about occupying a niche, do not put a lot of care into their young, have short life spans, and have evolved without significant competition for resources. They inhabit unpredictable environments and so invest in having many offspring. Many rodents and insects are R-selected species.

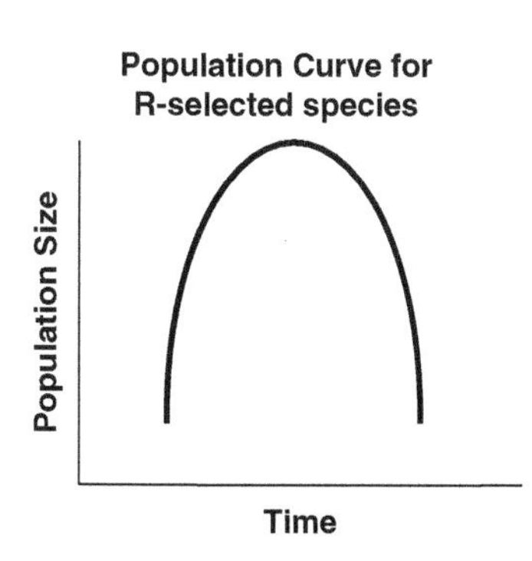

The R-selected growth curve resembles a hyperbole with a sharp increase called the **biotic potential**. When a species has reached its biotic potential, environmental conditions are ripe for exponential population growth.

The population levels off when it has reached its **carrying capacity**, meaning it has reached the maximum number of individuals that the environment can support. At this point, environmental resistance is pushing back on population growth decelerating the growth curve. After it has reached carrying capacity, overpopulation, lack of food, and resources eventually become problematic to the population, drastically reducing its population size.

K-selected species have longer life cycles, more competition for resources, inhabit more stable environments, fewer offspring and mothers who put more care into raising their young. Elephants, humans, and other primates are examples of K-selected species.

Because K-selected species have fewer offspring and encounter considerable competition for resources, their net population growth rate is typically slower than R-selected species.

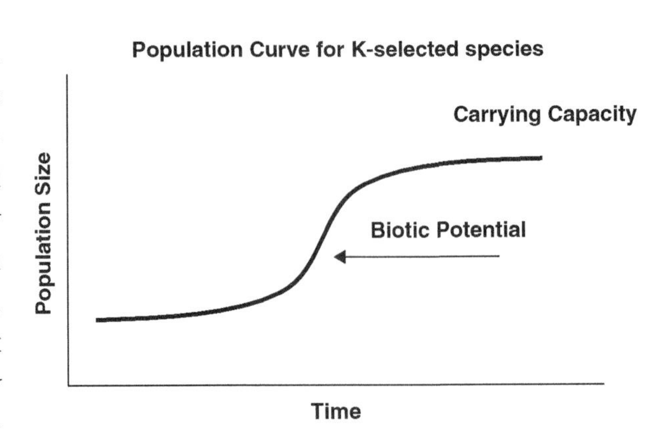

Once a K-selected species has reached carrying capacity, it is predicted that the population is maintained without experiencing a severe drop, as has been observed with R-selected species. This explains why the overall shape of the population growth curve is sigmoidal. Many demographers, scientists, and economists debate what the carrying capacity is for humans. Many argue that earth had already reached carrying capacity at 1 billion people while some argue that the planet can sustain as many as 50 billion, provided we learn to distribute food and assimilate waste more efficiently. (Consider that 20% of the world's population uses 80% of its resources!) Researcher and statistician Hans Rosling of Sweden predicts that the carrying capacity will level off at 10 billion, and we therefore must prepare for resources to accommodate that population number. Dr. Rosling also predicts that the average TFR worldwide will remain 2.[ii]

Most species cycle between both R and K growth curves. Humans, for example, are not without our devastating population controls. Consider such catastrophic episodes as the Bubonic Plague of 14th Century Europe, The influenza pandemic, The Smallpox epidemic, Ireland's Potato Famine of the 1800s, and the present-day AIDS epidemic. As a result of such dramatic loss of life, our population growth curve at times resembles more of an R-selected species than a K-selected species. All species meet with disease or some other form of population control.

Survivorship Trends

Just as population growth curves provide us with an abundance of information regarding a population, the overall age distribution of any population (whether it a human population or a wildlife population) tells a story and provides invaluable information about the population itself. Scientists must study population trends such as age distribution (as well as other trends such as ecology, population density, and reproduction), in order to effectively manage wildlife. Age distribution can address such matters as overall quality of life, quality of health, environmental stressors, and whether the population has been plagued by disease in recent years.

Consider these contextual examples: On a given island, given a specific species of turtle, why are there no 6-year old turtles? It may suggest that fecundity was low that year. In countries such as Japan, why are there inordinately high numbers of senior citizens? It may suggest overall good quality of life and health care.

Biologists who study population dynamics keep track of age distribution through a bookkeeping practice called **life tables**. Life tables highlight and display different variables and the age distribution of a given population, and often predict the age of death.

Individuals of a population born at the same time are called a **cohort**. If the population is of interest and researched by field biologists, the cohorts are often tagged in order to collect and analyze data on them. Biologists will note the overall age distribution of a population, (especially the age of death), and add this data to the life tables.

Survivorship curves are graphs that indicate age of death only (whereas life tables may include more information). There are three major types or trends of survivorship curves (see below).

Type I survivorship curves include sheep, zoo animals, and humans. This survivorship trend is common among K-selected species. In this type of species, the highest death rate is observed among the elderly of the population. Death rate of the young is relatively low.

Type II survivorship curves indicate a consistent death rate throughout the populations' life cycle. Death rate is the same in juveniles as it is among the elderly. Examples include many bird species such as sparrows and robins, turtles, and humans living in tough environments (where there is violence and famine, for example).

And finally, Type III survivorship curves indicate that death is highest among the populations' juveniles. Examples include plant seeds and many fish species. Fish eggs are often eaten by predators so juvenile death rate is high. This survivorship trend is common among R-selected species. See the three types of survivorship trends right:

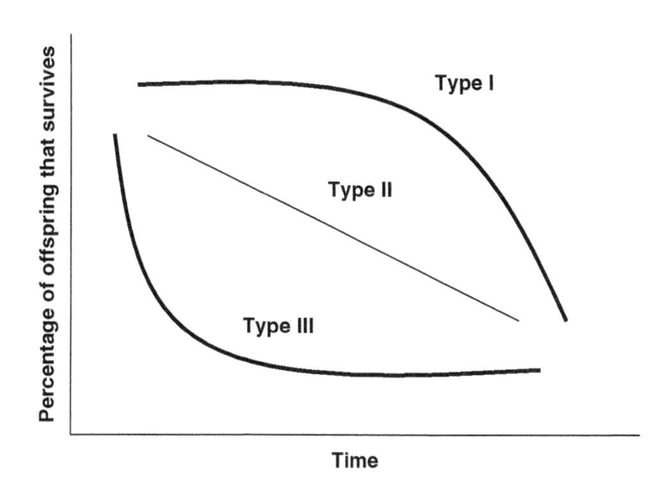

Review Questions

1. Define coevolution. Provide some examples.

2. Can you think of other examples of a Flagship species? What about an umbrella species?

3. Why are pollinators considered a "keystone species?"

4. Provide some examples of defenses that plants use to protect them against pests and herbivores.

5. Define symbioses. Define commensalism, amensalism, parasitism, and mutualism. Provide an example of each.

6. What are two possible scenarios that may occur when two species occupy the same niche?

7. How does phosphorus enter the ecosystem? How does Nitrogen, Phosphorus and Carbon Carbon enter the ecosystem? How are Nitrogen and Carbon released back into the atmosphere?

8. Define "trophic level." Please describe each trophic level.

9. Distinguish between primary, secondary, and tertiary consumers. How do the impact the energy flow of an ecosystem?

10. What is the difference between intersexual and intrasexual selection?

Discussion Questions for Class

A. Why are top predators and pollinators often considered keystone species?

B. How does human consumption affect the hydrologic, phosphorus, nitrogen, and carbon cycles?

C. Why are freshwater fish threatened with a higher rate of extinction than marine fish?

D. What are some effective adaptations of a predator? Prey?

E. How might parasitism promote genetic variability in a population?

F. How is predation beneficial to a population of prey?

G. Why is group living beneficial to species and prey? When is it a detriment?

H. How can we personally reduce our negative impacts on ecosystems?

I. Discuss some community interactions you would find in an ecosystem.

J. Define coevolution. Provide some examples.

K. Can you think of other examples of a Flagship species? What about an umbrella species?

Your Personal Sustainability Journal

Create or buy your own hummingbird feeder or regular bird feeder. Take note of how many visitors you receive.

Turn off water while you are brushing your teeth.

Recycle magazines and newspapers.

Take your batteries to the local organic market to be recycled (yes they have recycling bins!)

Buy organic cotton sheets and drapes.

Add your own suggestions here.

Endnotes

i. Jake J. Beaulieu, Tonya DelSontro, John A. Downing. Eutrophication will increase methane emissions from lakes and impoundments during the 21st century. Nature Communications, 2019; 10 (1) DOI: 10.1038/s41467-019-09100-5.

ii. For more resources on this please see http://www.ted.com/speakers/hans_rosling

Chapter 5
Nonrenewable Energy

DID YOU KNOW?

Earth Day in the Gulf of Mexico, April 20, 2010, is when a deadly explosion occurred at the Deep Horizon Drilling Rig, releasing what would eventually become 5 million barrels and 206,000,000 gallons of crude oil into the Gulf of Mexico. You may be familiar with this event as the BP Oil Spill. Eleven people died from the explosion and the wildlife casualties—thousands of birds, fish, whales, and endangered sea turtles among many others—are staggering. The disastrous effects to the food web have been incalculable. Disturbing photographs of oil-soaked brown pelicans, formally a federally endangered species, were splashed against news sources as a macabre symbol of the devastation. Are deadly accidents such as the BP Oil Spill worth our dependency on oil as our major energy source?

In this chapter, we will be exploring different nonrenewable sources of energy such as oil and other fossil fuels. We will also be examining their pros and cons and the multiple issues associated with each. And while reading this Chapter, please keep in mind that fossil fuels used for transportation accounts for 13% of global greenhouse gas emissions while energy supply accounts for 26%.[i] Please try to examine how your everyday activities require energy and fossil fuel use and what you can do to reduce your carbon energy footprint.

Fossil Fuels

Fossil fuels have powered the United States for over a century and have been integral in the advancement of technology, health care, and modernization. They have also been profoundly instrumental in increasing the human carrying capacity. Fossil fuels have undoubtedly provided great value to economies, industries, culture, personal health, and individual interests.

However, there is a downside to using fossil fuel with which we are becoming increasing aware; especially as its use continues, expands, and evolves. As a growing population continues to place incessant demand on fossil fuels, it is becoming difficult to ignore the adverse effects of its widespread use. To put it bluntly, fossil fuels are dirty and cause profound negative impacts to the environment, wildlife, and on human health. They create air pollution, water pollution and are a major contributing factor to global warming.

We will be closely examining both the pros and cons of using fossil fuels and will also take an in-depth look at their impacts on human and environmental health.

Fossil fuels were formed millions of years ago as the by-product of either mostly dead plant matter (that would eventually become coal) or dead plant and animal matter (that would, for the most part, eventually become oil and natural gas). The remains of these living organisms have been exposed to excess heat and pressure over the years, forming what we recognize as coal and petroleum. Because these remains were once living organisms, coal and petroleum contain many hydrocarbons (carbon and hydrogen covalently linked into long molecules) as well as traces of sulfur, nitrogen, and other impurities. Hydrocarbons are found in liquid, solid, and gas form, as we will examine in subsequent paragraphs.

Cheryl Casey/Shutterstock.com

Fossil fuels are **nonrenewable** sources of energy because there is a finite amount of them on earth. Once their supply is used up, they cannot be regenerated in the amount of time necessary to facilitate our energy needs and consumption. Despite this, they account for 85% of all fuel use in the United States. The United States owns 3% of the world's oil reserves, but due to our high transportation demands (two-thirds of oil goes toward transportation), we consume a staggering 30% of it. Currently, between 50% and 70% of crude-oil supplies have been depleted. Half of the remaining reserves are in the Middle East.

Petroleum

Petroleum is a mass of hydrocarbons found in a *liquid* form. These hydrocarbons are the remains of ancient plants, animals, marine organisms, and other organic material buried and trapped over thousands of years. Their remains were exposed to very little oxygen, which resulted in slow decomposition. This process forms crude oil, or petroleum—oil with impurities. We use this energy source for heating, transportation, and creating plastic and petroleum jelly, paraffin, and other products.

Crude oil is found within pores inside rock. The drilling and pumping process draws out the oil. **Primary production** occurs during the initial drilling stages. Layers upon layer of rock put immense pressure on compressed oil, so during primary production, when the pressure is released through drilling, a large quantity of oil bubbles to the surface. Over time as the pressure starts to weaken, the oil flow ebbs and other more extensive measures must be implemented to extract oil. Drilling resumes, but now deeper into the ground using more energy and exertion, sometimes assisted by adding gas to expand and increase pressure, causing more oil to bubble to the surface. This stage is called **secondary production**.

iurii/Shutterstock.comOffshore Drilling

Crude oil is then shipped, via truck, rail, or pipeline, to refineries where it is distilled at a fractionating tower and purified to eventually become refined oil. Since our energy economy and our transportation are dependent mostly on oil, incentive is high to make this process efficient, so distillation technology is state-of-the-art. Oil is finally transported to its destination.

In the United States, prices of gas at the pump are artificially cheap due to large government subsidies. This means that gas prices do not contain environmental cleanup costs and all other adverse side effects from oil dependency. If you have traveled to Europe lately, you may have noticed that gas prices are almost three times as high as U.S. prices. Europe implements **full-cost pricing**, meaning gas prices include costs of cleaning up air pollution, oil spills, and all other side effects to the environment from oil. Some argue that if oil were more costly, or in other words, we were paying its "full" price up front; then we would use it more sparingly and look for alternatives.

Toman1111/Shutterstock.com

Oil Refinery

The Middle East contains the largest oil and natural gas reserves in the world and we rely on them for a large proportion of our oil. Being dependent on other nations for oil decentralizes our energy economy. Should international relations become tense with nations of **OPEC** for any reason, then the United States and any country dependent on them would be (and have been) in a precarious situation. Some Middle Eastern nations are not our allies so being reliant on them for our oil supply may cause unnecessary friction in international relations, putting our national security in jeopardy. Along with the many adverse environmental impacts, national security is another compelling argument to decrease our dependency on oil. It also leads to oil price instability, which

Pros of Using Oil	Cons of Using Oil
– Include: it is relatively cheap and readily available to meet our transportation and energy needs – we have an ample supply for 50 more years or so... – it is energy efficient – it is easy to transport from country to country – distillation technology is advanced.	– Include: national security is compromised – it is a nonrenewable resource that we cannot rely upon much longer – oil spills – oil prices are volatile and unpredictable, which can harm the economy – due to depleting crude-oil supplies, we are forced to resort to obscure sources such as tar sands containing **bitumen**, a semisolid and less pure form of petroleum – during secondary production, sediment accumulates in crude oil, making it more difficult to distill – drilling may cause subsistence and other subterranean abnormalities – large government subsidies (that we end up paying for) are costly – carbon emissions and other by-products produce air pollution – petroleum systems and natural gas are responsible for 29% of methane emissions in the United States[ii] – air pollution created through oil use has profoundly negative impacts to human health – developing nations that happen to be rich in oil supply can, and have been, exploited by wealthy nations looking to expand their drilling – energy economy is centralized.

is not favorable for the economy. Finally, at the rate we are consuming oil, it is expected that we have ample supply to last for the next 50–100 years. This number is an informed, but imprecise estimate, as records and information containing relevant data are unreliable.

Natural Gas

Found on coal beds, natural gases are hydrocarbons found in their gaseous form, mostly methane (CH_4), but sometimes natural gas is in the form of propane (C_3H_{12}) and butane (C_4H_{10}). Natural gas is the product of decomposed organic matter (such as marine microorganisms) that lived millions of years ago. This decaying matter eventually settled to the sea floor and became buried with sand, silt, and mud. In the presence of anaerobic conditions and an abundance of heat and pressure, the dead organic matter eventually turned into a hydrocarbon gas (methane, butane, or propane), or what we know as *natural gas*.

Because of the specific and controlled way natural gas was formed, it is often covered by porous sedimentary rock. Natural gas is low in density, and if rock is suitably porous, gas will naturally rise up out of the rock. *Conventional natural gas deposits* are associated with oil and make up 46% of natural gas resources. The rest—*unconventional natural gas deposits*—is associated with oil shale and coalbeds.[iii] Russia has the largest natural gas reserves, possessing five times that of the United States. Other countries with considerable natural gas reserves include: Iran, Qatar, Saudi Arabia, and Venezuela.[iv]

Natural gas is volatile and easily combustible, making it a valuable energy source. It is often harvested in a process called **fracking**. Fracking involves using water, sand, and chemicals to break up rock, build up pressure, and release natural gas from rock beds. It has allowed natural gas companies to access large quantities of natural gas in a relatively short amount of time. This is a controversial process since the chemicals used are potentially toxic, and therefore threaten to contaminate our land, air, and water supply. Furthermore, according to the USGS, fracking has on occasion contributed to earthquake activity.[v]

After it is pipelined to the power plant, natural gas is often heated to boil water and the steam is used by turbines to generate electricity. Natural gas contains impurities and must be refined to remove acid gases, such as hydrogen sulfide and carbon dioxide, and other toxic impurities such as mercury and cadmium.

The U.S. Department of Energy is currently conducting research to examine how to utilize microbes for converting natural gas into a liquid form—without the input of more energy. This would ultimately lessen our reliance on petroleum.[vi]

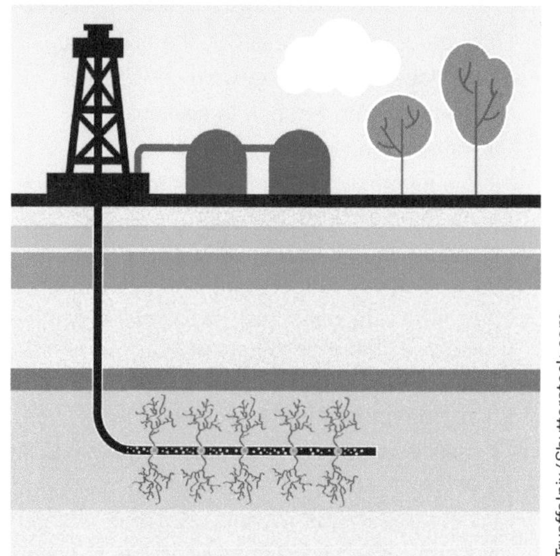

Hydraulic Fracturing is the Fracturing of Rock by A Pressurized Liquid

Natural Gas in Landfills

Landfills are the third largest source of human-generated methane gas in the United States.[vii] The Environmental Protection Agency (EPA) has established an outreach program called Landfill Methane Outreach Program (LMOP) that recovers and reuses landfill gases—like methane—for energy. Methane is captured through a vacuum-like device and converted to electricity. This electricity can be sold to the power grid, used on-site, or used to fuel power plants, vehicles, homes, and neighboring towns. It is also used as a means for **cogeneration** (combined heat and water) in which heat is captured and used to produce steam.[viii]

The benefits of this outreach system are many. Currently 48 states use the program for heat, fuel, gas and electricity. It creates jobs through the marketing, design, implementation, and operation of the program. It also results in an overall decrease in greenhouse gas emissions because it uses up the methane instead of allowing it to descend into the atmosphere where it would ultimately act as a greenhouse gas. It would also theoretically improve the overall air quality by reducing the need for fracking or other fossil-fuel sources (coal and oil) which produce air pollutants such as sulfuric dioxide, nitrogen oxides, and particulates.

Critics of the Landfill Initiative are concerned about the amount of methane produced—which is a much more potent gas than carbon dioxide (20 times stronger than carbon dioxide at trapping heat over a 100-year period and 84 more times potent in a 20-year period). Since only a small fraction of the methane is utilized, much of it is released into the atmosphere or burned on-site. In fact, in the United States, landfills are responsible for 28% of methane emissions.[ix] This makes it hard to reconcile generating and using methane as an energy source. Critics say that perhaps we should be focusing on cutting down out amount of waste altogether and turning clean-burning renewable energy.

Pros of Using Natural Gas	Cons of Using Natural Gas
– Include: it releases less CO_2 than oil or coal – it provides a high energy yield – a relatively ample supply is available – it is cheap and easily transported – it can be renewable if microbes in landfills are used to break down biomass – it can be used for cogeneration – because EPA standards are increasing, coal power plants have become more expensive to run, so there is a shift toward natural gas – fracking allows domestic fuel to be much cheaper than from other foreign sources – it uses less water and creates less water pollution than coal plants – it has greater flexibility with the power grid – it has given the economy a boost as we have shifted to a cheaper means of energy.	– Include: carbon emissions cause pollution – government subsidies are costly – it releases methane (a greenhouse gas) – fracking uses toxic chemicals and so causes air, water, and land pollution – fracking is also linked to earthquake activity – natural gas prices are volatile and may lead to a higher cost for electricity. This may harm the economy and, in extreme cases, cause industry to shift back to cheaper coal plants – Natural gas leaks methane gas, which is much more potent at trapping heat than carbon dioxide. Petroleum systems and natural gas are responsible for 29% of methane emissions in the United States[x].

Oil Shale

Unlike the other forms of fossil fuel, oil shale is still trapped in its original rock formation, formed from the accumulation of clay, silt, mud, and organic matter at sea beds. Oil shale is composed of compressed sedimentary rock that contains, not oil as the name suggests, but organic matter called kerogen. Unlike coal and oil that comes from dead plant and animal matter, oil shale is derived from decomposed cyanobacteria, marine algae, and other microbes. When it is heated, it releases gaseous kerogens that are combustible and may be used as an energy source. Oil shale garnered attention as we started looking for alternative fuels, but research was stalled due to the excessive energy used during extraction. In addition, little is known thus far about the environmental impacts of using oil shale as an energy source.

Pros of Using Oil Shale	Cons of Using Oil Shale
– Include: it provides the benefits of using fossil fuels, but in a much shorter amount of time – it exists in ample supply – it is cleaner burning than other fossil fuels – advanced horizontal drilling and fracking allow access to large volumes of oil shale.	– Include: it is expensive – produces a large environmental impact such as air, water, and land pollution – experts are still unsure of how to deal with waste disposal.

Coal

Fun Fact

Oak was the major fuel source of medieval Europe. Europeans also used oak to build temples, shrines, and eventually ships during the exploring years. Deforestation rates further escalated as oak forests were rampantly cut down for agricultural use. Inevitably, Europe faced an oak shortage that forced them to start looking for alternatives. The 1700s discovery that coal was an efficient fuel source was the solution to their energy crisis.

Coal has been used since ancient times; it was used by the Hopi Indians of the Southwestern United States as well as the Romans in the third century. But it was the recent revelation that coal could burn hotter, cleaner, and release more energy than wood that propelled coal technological advancement in the 1700s. It solved the European energy crisis as well as saved the rest of oak forests and pine forests of the Baltics and North America (a favored solution) from deforestation. This technological leap eventually paved the way for the industrial revolution.

Coal is a *solid* form of hydrocarbons that are the remains of plant matter. It was the major energy source of the Industrial Revolution of the 19th Century when pulverized coal was placed into coal power plants to produce energy. Today, coal is pulverized and burned in a furnace at coal power plants. Water runs through the furnace through a piping system, and then heated by means of the combustion process. Water is heated until it boils and creates steam, which is pushed through a turbine and produces electricity. The water cools and condenses and is placed back into the furnace.

Coal is currently one of the major sources of energy in the United States and worldwide. The Unites States' largest coal reserves are found in Wyoming and Appalachia. It contains many chemical impurities (noncarbon elements) depending on the type of coal (lignite, bituminous, anthracite—in order of increasing carbon purity); most coal also has hydrogen, nitrogen, oxygen, and sulfur (impurities) within its structure. Technology has progressed so as to reduce the amount of sulfur and nitrogen in coal as well as to limit its carbon emissions. Refining coal is not as intensive as refining petroleum. It is treated with water or chemical bath to remove impurities. This process can remove up to 30% of the sulfur content and reduces the amount of acid rain and other negative impacts that the combustion of fossil fuel has on the environment.

Madlen/Shutterstock.com

Coal seams are found either close to the earth or deep underground. Conventionally, coal is accessed by drilling into mountains and getting into greater depths by using explosives, which create major blasts, dispersing rock and debris. **Overburden** (rocky waste) is removed until a layer of coal is reached. This is called **surface mining** and is sometimes called *strip mining*. Often times, this entire process includes taking off the whole top of a mountain! The removal of overburden can be repeated several times or as needed to expose coal. The copious amounts of rubble and rocky waste are piled on to accompanying mountain valleys, causing lots of pollution and disturbance. The overburden contains toxic elements such as cadmium, lead, and mercury that can get absorbed into the water table, potentially and often contaminating it. Furthermore, sulfuric acid produced by this process can wash into the soil and water supply, resulting in **acid mine drainage**, a particularly toxic threat to the environment.

Another type of surface mining is **contour mining**, which is performed when coal seams are exposed on the mountain side. These mines are smaller but still generate considerable waste. **Pit mines** are created when when coal seams are deep in the mountain and deep pits must be dug to access coal.

Jim Parkin/Shutterstock.com

Surface Mining

Another type of mining is called **subsurface mining**. This form of mining is implemented when minerals are found deep in the ground, and mining usually carried out through the use of underground shafts. Sometimes tunnels are created to facilitate subsurface mining or even large rooms with pillars to support a shaky roof. Loose coal is collected and put on a conveyor. Sometimes dynamite is used to release coal. Subsurface mining comes with many risks: possible death or injury; and sink holes, or subsidence in the land.

An alternative to conventional mining is biomining, when microbes are used **in situ** to break down overburden and expose valuable minerals. This could potentially reduce the amount of waste associated with mining.

farbled/Shutterstock.com

Subsurface Mining

Pros to Using Coal	Cons to Using Coal
- Include: it is energy efficient - coal emits less pollution and CO_2 than oil - it is the cheapest to extract - it is safer and easier to ship - there is 10 times as much of it as natural gas and oil - we have an ample supply that could supply us with energy for a few hundred more years - the EPA has enforced stricter regulations on sulfur dioxide, nitrogen oxide, mercury, and particulates, producing less pollution and cleaner coal.	- Include: land disturbance and pollution is extensive, the extensive pollution and disturbance can destroy forest habitats and pollute streams - coal use places high demand on water resources - it releases sulfur dioxide that is linked to acid rain - it releases nitrogen oxide which is linked to acid rain - it releases carbon dioxide that is believed to be the leading cause of global warming - it releases carbon monoxide, a colorless, odorless toxic and often deadly gas - it releases toxic chemicals such as mercury and cadmium - it releases trace amount of uranium, which can produce radioactive emissions - it is responsible for 10% of methane emissions in the United States[xi] - it produces sink holes or subsidence and acid mine drainage - it causes black lung among other ailments, as mining practices take a human toll - carbon emissions can be catastrophic - because the EPA has enforces stricter regulations on sulfur dioxide, nitrogen oxide, mercury, and particulates, coal plants are more expensive to run.

The Impact of Fossil Fuel on Biodiversity and Human Health

We examined various pros and cons of fossil fuels above, but let us get a little more in-depth by examining how the use of fossil fuel specifically impacts ecosystem biodiversity and human health: As we mentioned above, the use of fossil fuels is linked to sulfur dioxide and nitrogen oxides that cause acid rain. This damages forests, buildings, and bodies of water. Fossil fuel use is linked to an increase in various types of air pollution which in turn can negatively impact agriculture and damage forests. This is especially damaging to primary growth forests which harbor a good amount of earth's biodiversity.

The use of fossil fuels has been linked to global climate change, which has many adverse effects on biodiversity, ecosystem function, and human health (See Chapter 12 for more information).

Using fossil fuels can lead to the risk of such catastrophic oil spills such as the BP oil spill mentioned in the opening paragraph of this chapter. Furthermore in 1989, the Exxon Valdez spilled 37,000 tons of crude oil in Prince William Sound in Alaska.[xii] This results in devastating loss of biodiversity. And the creation and expansion of oil fields encroaches on fragile ecosystems such as Tropical Rainforests and some that are typically used for conservation—such as the Arctic.

Fossil fuels produce particulate matter like soot smoke and dust. This air pollution may impact human health (see Chapter 14) by increasing the risk of respiratory complications such as bronchitis and asthma, which can lead to long-term complications such as CPOD.

Nuclear Energy

In 1986, in Chernobyl, Ukraine (then, the Soviet Union), several explosions destroyed a nuclear reactor, releasing radioactive material, and contaminating the area for an estimated 100 years. As a result of such devastation, locals cannot drink the water or eat locally raised crops or meat from livestock. Dramatic increases in both cancer and birth defects have also plagued the area. Nuclear energy is a nexus of global controversy, not only because of the risk of accidents and subsequent health risks, but also because nobody can seem to agree how to effectively dispose of nuclear waste. Is the potential for accidents such as Fukushima (where the safety systems failed), Three Mile Island, and the Chernobyl disaster worth a dependency on nuclear energy? If so, what measures can we take to make the process safer and ensure public safety?

Nuclear energy provides 8% of the world's energy needs. France receives 80% of their electricity from nuclear power and the United States receives approximately 10% from nuclear power. Nuclear energy does not come from fossil fuel and is not produced from hydrocarbons. Rather, it is harvested from a process called **nuclear fission**. Nuclear fission produces 10 million times more energy than traditional energy sources such as coal or oil! This process involves the splitting of an atom, a radioactive element, such as uranium, with a heavy molecular weight that decays to become a radioactive **isotope**. This process is exothermic because it emits heat energy which is used to heat water, create steam, and generate electricity. The fission reaction must occur at a very specific rate, and must be stopped or adjusted when necessary. Large quantities of water are used in this process not only for steam generation but to act as a coolant as well. Water pressure and temperature must be heavily controlled to prevent a meltdown.

Uranium is collected by either surface mining or subsurface mining. After the mining process, it must be enriched to increase the concentration of the unstable isotope, U-235. Later, it is made into pellets and placed into fuel rods. Actual nuclear fission takes place in the reactor core—where the turbines generate electricity and condensers cool steam.

The average life span of a nuclear power plant is roughly 50 years. When they have reached the end of their run, nuclear power plants cannot be abandoned, they must be **decommissioned**. Radioactive waste must be carefully removed and placed in a storage facility for at least 100 years, then monitored for at least a thousand years to prevent radioactive leaks.

In 1982, the U.S. Nuclear Waste Policy Act was passed, positioning the Department of Energy in charge of nuclear waste disposal. The ideal site for disposal, according to The Act, must be geologically stable and located at a safe distance from groundwater. The site must be able to hold waste for tens of thousands of years, or at least until the waste is no longer radioactive. Currently the United States is stalemate concerning how to dispose of its radioactive waste (which is piling up at power plants). The consensus was that we would store our waste underground at Yucca Mountain in Nevada. As scientists started learning more about this landscape and the multiple risks involved, not to mention the resistance from concerned citizens, the project stalled and has yet to move forward.

F. ENOT/Shutterstock.com

Pros of Nuclear Energy	Cons of Nuclear Energy
– Include: carbon emissions are low – a vast energy supply is conveniently produced in such a small amount of material, 1 kilogram of uranium produces similar amounts of energy as 100,000 kilograms of coal – land remains relatively undisturbed – without accidents, it is very safe – security and safety measures have increased significantly since September 11, 2001 – the technology is already established – some consider it green energy because it does not release greenhouse gases and so is not directly linked to global warming.	– Include: lending institutions (for construction purposes) need loan guarantees because they are so expensive to build and risks are high – power plants are more expensive to build than coal power plants – there is great disturbance and water contamination associated with mining for Uranium – Nuclear power plants are vulnerable to natural disasters – knowledge use by terrorists would be devastating – accidents and spills prove to be cataclysmic – no clear solution on how to dispose of waste is evident – the net energy yield is low compared to coal – the risk of reprocessing or removing plutonium from nuclear waste could cause a security threat since it is used to build nuclear weapons – concerned scientists and critics believe that the NRC (Nuclear Regulatory Commission—the agency that oversees safety of the power plants) is too lenient and therefore ineffective – some argue that the standards for relicensing are too loose – tritium (radioactive hydrogen) has been known to leak from nuclear reactors and into the water table (tritium is linked to cancer) – some argue that the risk/reward system is distorted since an accident or nuclear meltdown can eradicate all benefits of using nuclear power. It is estimated that the costs to clean up Chernobyl, for example, far outweighed any economic benefits.

The Impact of Nuclear Energy on Biodiversity and Human Health

The construction of nuclear power plants releases small amounts of greenhouse gases during its initial construction, so impacts would include most of the ecological impacts of using fossil fuels; Water used to cool the reactors is dramatically above ambient temperatures. This may accelerate or compound the ecological impacts of climate extremes that we will examine in Chapter 12, and people and wildlife exposed to radiation can suffer disease and genetic changes, and in extreme cases death. Tritium leaks and exposure to tritium can lead to cancer.

On a lighter note, because of the significant risks associated with nuclear power plants, the surrounding area usually serves as protected areas for wildlife conservation and research. Provided there are no radiation leaks or meltdowns, this would be beneficial to the ecosystem and advancements in science.

How Do We Make Nuclear Energy Safer?

Let us examine some final thoughts on how to make nuclear energy safer for the general public. Some possible course of action may include enforcing strict earthquake and fire regulations for nuclear power plants, especially for older and aging facilities. We also need to design a solid plan of action in the case of disasters like floods—especially in coastal areas. We should determine a safety storage plan once and for all and remove the accumulating waste from the facilities, and devise rigorous security plans to make these power plants less vulnerable to terrorism. And finally, the NRC must enforce stricter regulations and their oversight should be consistent and thorough.

Are You a NIMBY?

Have you ever heard or uttered the phrase, "Yes, but not in my backyard?" (**N**ot **I**n **M**y **B**ack**Y**ard?) If you have, or have supported this notion in any way, then you might be what people may refer to as a '**NIMBY**.'

What exactly does this mean? Often times, people will support a particular cause or action, even though there may be possible environmental or health risks associated with it, because the benefits seemingly outweigh the costs.

For instance, a person may be pro-Nuclear Energy, or supportive of a pipeline that transfers oil, or in favor of the concept of hydraulic fracking. They may back all of these as possible purveyors of energy—UNLESS the nuclear reactor is in *their* town, unless the pipeline goes through *their* backyard, unless the hydraulic fracking happens in *their* community. It is acceptable to carry on as long as it does not affect their families directly, or the health of their family directly. If these are in fact happening 'in your backyard,' or when one's family is directly affected, the costs then outweigh the benefits. In other words—"not in my backyard."

If it is health or environmental risks that people object to affecting them in the "not in my backyard" scenario, then perhaps it is time to take an honest assessment of the situation. Concede to the notion that it will, in fact, affect another person's family. Is this fair and just? Why is it acceptable to subject another family to the same risks that you or your family is adamantly opposed to? If you take the, "Yes but not in my backyard" approach, then perhaps in fairness you should reconsider your stance on the matter (as long as it is a real risk and not just a cosmetic concern).

Case Study

Oil Companies in Ecuador

Dr. Morley Read/Shutterstock.com

Ecuador Rainforest

Olgysha/Shutterstock.com

Giant Tortoise on Galapagos

Ecuador is a picturesque South American country located on the earth's equator. It is home to the Galapagos Islands; rich with biodiversity, famous for giant tortoises and endemic species like marine iguanas and a variety of finch species. It is also home to the Andes Mountains that are endowed with culture, hot springs, mountain villages, and sporadic volcanic activity. The Amazon Jungle is found nestled in the interior of the country.

Ecuador is also rich in oil supply and happens to be an OPEC country. Present day, oil is Ecuador's number one export, followed by bananas, shrimp, cacao, coffee, and tourism. Oil exploration in the Jungle of Ecuador began in 1950—commencing with seismic activity (explosives). In the year 1960, the first oil wells were drilled. From this point forward, oil companies expanded their quest throughout the Amazon Jungle in search of more oil.

Oil exploration has resulted in the construction of cities in the otherwise pristine Jungle, as there is great monetary incentive for people to relocate and work in what was once uninhabited land but now rich with oil. Currently, the target areas for oil drilling are home to Indigenous tribes that live in isolation (such as the Tagaeri Taromenane and the Waoranis).

This is an obvious reason to pause and develop a deeper understanding on the gravity of what is happening below the surface in the jungles of Ecuador. This is a multi-tiered issue that is complex and carries with it many environmental, social, and socioeconomic considerations.

(Continued)

Let us explore some pros and cons:

Some possible pros and arguments *in favor* of oil exploration in Ecuador include the generation of money and revenue for public interests such as roads, hospitals, schools in rural areas, and academic scholarships for higher education; the creation of jobs and subsidies for the elderly who need monetary assistance. Because oil is so lucrative and brings money in to the economy.

There are many cons to the issue that need to be addressed as well. Such cons fall into both the environmental and social considerations. Let us first explore the environmental cons; oil exploration causes a lot of air pollution because the natural gas is burned all year long in many locations. The vehicles and machinery used cause major land disturbance; ground pollution is a result of oil extraction and oil spills that occur during different stages of the extraction process (extraction, transport, and waste) and the disposal by oil companies often does not always use these safety measures; perforation water, which is reinjected into the ground, contains a lot of toxins that pollute water and contaminate drinking water that plants and animals depend on. The creation of roads is one of the main impacts because access roads invite illegal loggers, illegal hunters and settlers who are attracted to the resources, causing big impacts to the environment and the local communities; Unsustainable agriculture (such as the African oil palm) is one of the main causes for the jungle being cut down for crops (enabled through the construction of access roads) causing an overall loss of biodiversity.

Negative social ramifications include: a loss of local cultures because their life is changing by exposure to material and alcohol—the closer natives are to roads and cities, the worse it is. In addition, health problems to local people have been increasing (possibly because the air and water is polluted causing respiratory problems, cancer, etc.) And finally, there is an observable change in traditional activities of the indigenous people. They have stopped hunting animals in favor of grazing animals such as chickens and pigs.

Review Questions

1. Why are coal, oil and natural gas considered "fossil fuels?" Why are they considered "dirty?"

2. Distinguish between primary production and secondary production of oil drilling.

3. Define: bitumen and fracking. Why are these controversial terms?

4. Distinguish between Surface mining, subsurface mining, and biomining.

5. Define: isotope, NRC, nuclear fission, and decommission. These terms are associated with *what* type of energy?

Discussion Questions for Class

A. How has the use of nonrenewable resources contributed to many of the environmental problems we face today?

B. Discuss the pros and cons of petroleum, natural gas, coal, and shale oil.

C. The OPEC (Organization of the Petroleum Exporting Countries) include Algeria, Angola, Ecuador, Iran, Iraq, Kuwait, Libya, Nigeria, Qatar, Saudi Arabia, United Arab Emirates, and Venezuela. What are the political ramifications of dealing with OPEC, some of which are known to have supported terrorism?

D. Should we move toward a more decentralized energy economy? A decentralized energy economy is one in which we rely on several different energy sources. What are benefits of such a shift?

E. How does using landfills as source of natural gas and electricity a benefit to the environment?

F. What lifestyle changes are necessary for us to conserve oil?

G. Europe in the 1700s looked to coal to answer their energy deficiency crisis. Are we currently facing a similar energy crisis? If so, what alternatives are we looking to? Are they effective?

H. A great deal of nuclear energy research took place in the United States in the first part of the 20th century and participating scientists believed that it would be a major energy source by the middle of that century. What happened for nuclear energy to lose its momentum? Consider disasters such as Chernobyl, Three Mile Island, Fukushima, and possible threats of terrorism.

Your Personal Sustainability Journal

There are ways you can take action online to ensure safer nuclear energy. Demand that Congress to have the NRC enforce regulations in a timely manner.

In regard to the concern of fracking: identify possible impacts of fracking in your area and distinguish what is reliable information versus propaganda. Communicate with policy makers, journalists, scientists, and community action groups to make this the forefront of community discussion.

Cut down on overall energy use.

Instead of driving, walk or ride your bike! Take public transportation. Carpool.

Cut down on the amount of plastic you use. Seek out alternatives like re-usable compact bags and re-useable utensils.

DWP will pick up your old fridges and replace them with energy-efficient fridges for free!!!!!! This is worth looking into!

Some states will pay you for your old car if it does not pass the smog check. It is worth looking into.

Line-dry clothes and skip the dryer!

Raise the temperature on your refrigerator and raise or lower the thermostat (depending on conditions) by 2°.

Get the oil changed in your car regularly and maintain tire pressure.

Reflect how many pounds of CO_2 you release each year through driving. Can you reduce this amount?

Educate yourself. Learn more about the different sources of energy and their environmental impacts. Help educate other people so they too can make informed decisions.

Add your own suggestions here.

Endnotes

i. IPCC 2007

ii. http://epa.gov/climatechange/ghgemissions/gases/ch4.html

iii. Energy Information Administration. 2012. Annual Energy Review. Table 4.1 Technically Recoverable Crude Oil and Natural Gas Resource Estimates, 2009.

iv. Energy Information Administration. 2012. International Energy Statistics. Proved Reserves of Natural Gas.

v. http://pubs.usgs.gov/of/2015/1070/

vi. www.energy.gov

vii. Epa.gov

viii. www.epa.gov

ix. http://epa.gov/climatechange/ghgemissions/gases/ch4.html

x. http://epa.gov/climatechange/ghgemissions/gases/ch4.html

xi. http://epa.gov/climatechange/ghgemissions/gases/ch4.html

xii. ITOPF (2006). Summaries of major tanker spills from 1967 to the present day. http://www.itopf.com/casehistories.html#exxonvaldez

Chapter 6
Renewable Energy

DID YOU KNOW?

In the northern hemisphere, the south-facing side of homes and buildings receives the most net sunlight throughout the day. Many homes today are being built to orient windows to the south-facing side, maximizing the amount of sunlight throughout the winter to capture the sun's radiant energy for warmth. These homes also deliberately have few east- and west-facing windows in order to shield the home from the sun in hot summer months. This clever architectural design is quickly gaining popularity and being used in order to save on escalating heating costs as well as participate in the green movement and reduce carbon emissions and energy bills.

Why Renewable Energy?

Renewable energy is defined as energy originating from a source whose supply is indefinite or renewed (almost) as quickly as it is used. Most renewable energy supply comes either directly or indirectly from the sun. Solar energy, as illustrated in the above example to heat homes, is one example of renewable energy, but in this section, we will be exploring several other sources of renewable energy as well.

Benefits of renewable energy can be categorized with political, economic, social, or environmental considerations. From a political standpoint, using renewable energy improves our national security by relying less on **OPEC nations**. Reducing our dependency on these countries might potentially improve national safety by avoiding unnecessary conflict. Using more renewable energy sources also decentralizes our energy economy and helps reduce our financial debt to oil-exporting countries. It creates new jobs and a new economy around renewable energy and its science, innovation, sales, technology, and marketing strategies. Renewable energy is reliable and is becoming more affordable with the availability of tax cuts, rebates, and reasonable financing options. It provides increasingly affordable energy with stable and predictable energy prices, which is beneficial for the economy. The shift is becoming socially popular and gaining momentum within the green movement.

Environmentally, it burns cleaner and reduces carbon emissions and pollution. Fossil fuel use has been linked to cancer, breathing problems, heart attacks, and neurological damage. Shifting away from fossil fuels and toward renewable energy therefore improves public health by improving overall environmental quality and potentially reducing overall healthcare costs.[i] And by using renewables, we do not have to worry about running out of energy supply. About 25% of the world's energy supply is renewable and 18% of the United States' energy supply is from renewable sources. Renewable energy is the fastest growing energy sector, with growth projected at 2.3 % a year. Most of this growth is in the form of solar and wind energy (Source eia.gov). Many communities that would like greener energy or more energy independence are moving toward **MUNICIPAL AGGREGATION or CCA's** - Community Choice Aggregation. Communities with CCAs choose their power supplier which gives them more control over energy sources and increases competition among energy suppliers.

Solar Energy

Sunlight does not take up space or have mass, so it is not considered **matter**. Rather, sunlight is classified as **electromagnetic energy** that can be captured, transmuted, and used for our energy needs. Some of its many uses include heating our homes and heating our water supply, as well as generating electricity. (**Photons** are light acting as matter-like particles; but for the sake of this chapter, we will consider light as having wave-like properties.)

Active and Passive Solar Energy

According to the legend, the Greek scientist Archimedes used a collection of solar reflectors to concentrate solar energy and use it to burn Roman ships during the 212 BC Battle of Syracuse. This feat has been reconstructed recreationally by volunteers and proven to be a possibly viable (albeit diabolical) endeavor. Historians are dubious of the historical accuracy regarding this event; however, it reminds us that there are many creative and innovative uses of solar energy.

Solar energy is growing, and it is growing rapidly as it becomes more affordable and accessible and efficient. In the United States, solar jobs have increased 123% since 2010, with over 200,000 workers (Source: National Solar Jobs Census).[iii,iv] Solar is not only providing energy for residential homes; many companies and industries are shifting toward rooftop solar cells to improve their reputation, company profile, marketing, and for reducing their operating costs. In 2017, PV accounted for 55% of newly installed energy capacity, more than nuclear energy and fossil fuels combined (Source: REN21).

Active solar energy is a process that uses conductors such as metal to trap sunlight. These conductors are usually a dark color, since dark colors absorb more wavelengths of light than lighter colors. This ensures that more electromagnetic energy is available for energy use. Sunlight is then transferred to a water supply, surrounding air, or pipes to heat homes. The heat may also be stored in a liquid form to be used later. We will be examining the different types of active solar energy in the following paragraphs.

Concentrated Solar Power (CSP) is a form of active solar energy that uses hundreds and sometimes thousands of mirrors to concentrate sunlight and amplify its reflective properties. A typical CSP farm is roughly the size of 200 football fields. The largest CSP plant is located outside Boulder City, NV, and was constructed in 2007.

In using CSP, a solar collector is used to convert sun's energy to steam through heating a mineral oil fluid, which is eventually transferred into electricity with use of steam turbines. Solar receivers track the sun's movement to maximize energy production. It is a relatively simple process, and because of its simplicity, it generates renewable energy in quantities that are increasingly becoming comparable to coal or other fossil fuels energy sources.

paulrommer/Shutterstock.com

Concentrated Solar Power

Another form of active solar energy is **photovoltaics**, often referred to as solar cells. Small-scale PV systems comprise the majority of solar installations and typically found on residential rooftops. Along with CSP, large-scale PV systems are a huge component of solar energy's generation of electricity. Photovoltaics use elements like silicon to capture sunlight and convert it into electricity right in the cell. During this process, solar energy energizes electrons on the silicon collector and the electrons flow to produce energy. The solar panels are connected to a circuit that provides a path for the flow for electrons, providing an electricity supply.

A concern for many people is how the amount of sunshine, or lack thereof, would impact the availability of solar energy. Is solar energy effective in places that do not experience an abundance of yearly sunshine? This is a reasonable concern, and it turns out that many solar systems are energy effective all over the globe, even where there is not an abundance of sunshine. A solar panel in Portland, Maine (which experiences long harsh winters) for example, would generate 95% as much power as the same system in Miami, and 85% as much power as the same system in Los Angeles.[v] This is a very generous amount of energy in comparison.

When businesses or homes generate more electricity than they use, often the electricity is placed back to the grid.

Shutterstock.com

PV Solar Cells

If you have ever stepped into a greenhouse on a sunny day, you are aware of the powerful energy of the sun. Greenhouses are an example of **passive solar energy**—or a means of harnessing sun light without the use of a collector or solar cell. Passive solar energy typically uses natural materials such as water, stone, and brick to store the sun's heat (and typically passive solar energy uses no moveable parts). The contextual example at the beginning of this section—houses architecturally designed to maximize sunlight—is another example of passive solar energy.

Iladyjane/Shutterstock.com

Pros of Using Solar Energy	Cons of Using Solar Energy
- Include: it lessens our dependency on oil and OPEC nations (many of whom support and fund terrorism) - it decentralizes our energy economy - it is renewable - it is not very costly - it creates jobs - it is convenient for those who live in rural areas, not in close proximity to the energy grid - solar panels are relatively easy to install - it has no carbon emission and little pollution - tax credits and rebates are available - reasonable financing is now available, PV rooftop installations are becoming more affordable and more cost efficient - unlike nuclear energy and coal, PV systems do not require water to generate electricity - CSP can store sun's energy as heat and may use it to make electricity even when the sun is not shining - price of solar energy is stable and predictable.	- Include: the energy transfer is not 100% efficient and so energy yield is not high - sunlight may not be available at peak times of need - fuel cells can be expensive to install in homes - solar farms that are constructed in the desert or other fragile ecosystems cause large disturbances and compromise wildlife - water is used to manufacture PV solar components - PV manufacturing involves some toxic chemicals such as sulfuric acid, nitric acid, and hydrochloric acid. If not disposed of or handled carefully, these chemicals could be harmful to human and environmental health - although there are no fossil fuels associated with the generation of electricity, there are fossil fuels associated with other stages of the solar energy life cycle, such as manufacturing, installation, maintenance, materials transportation, decommissioning, and dismantling - solar farms often produce more electricity than is needed at off-peak energy times, and the U.S. power grid is limited in its storage capacity - solar collectors use mirrors that focus sunlight onto a receiver. This focused concentration of heat in the desert can accidently cause burns to birds and other wildlife.

Wind Energy

The sun's unequal heating of earth's surface causes a difference in pressure gradients which we experience as wind. Wind energy has been used by humans for 2000 years for various purposes such as grinding grain and pumping water. In the United States, between 1870 and 1930, farmers used wind energy to pump water. With the arrival of the "New Deal" in the early part of the 20th century, windmills phased out, and grid-connected electricity was introduced to the countryside and rural areas. Wind energy in Europe has been more long term and consistent than that of the United States.

Wind energy is one of the cleanest, cheapest forms of energy, and is one of the fastest growing sources of electricity in the entire world. In the year 2000, there was 2.53 GW of wind energy installed across the US. In 2013, it was 60.72, and 404.25 is projected for the year 2050. The state of California is currently working toward its goal to make 100% renewable by the year 2045. of its energy renewable

majeczka/Shutterstock.com

by the year. To reach this goal, Southern California Edison is overseeing and making contracts with private construction companies erecting wind farms in the desert to connect to its current electrical system, providing neighboring metro areas with renewable wind energy.

Wind farms are comprised of wind turbines. Wind turbines have rotating blades, or sails, that when rotate, produce an electrical current that is transformed into electricity. Turbines are either horizontal-axis turbines or vertical-axis turbines. They can vary on size depending on whether they meet residential needs or are part of a wind farm. Off shore turbines tend to be large and generate the most power. Turbines typically have three parts: the *sails*, the *tower*, and the box behind the blade called the *nacelle*. The nacelle is where motion is transformed into electricity. Transmission lines carry the electricity to the power grids close to the urban areas where energy needs are high. Wind farms are typically built where there are high wind speeds—such as the desert or in natural wind tunnels. Unfortunately, wind tunnels are also often wildlife corridors, passages that connect one parcel of a species' land with a contiguous parcel of land. This can disrupt wildlife corridors and species' range patterns. Wind farms are more productive in oceans where wind speeds are far greater than that on land.

Andrzej Wilusz/Shutterstock.com

Wind turbine

But off-shore wind farms are not without controversy. High-value property owners do not want their ocean view compromised by off-shore wind farms.

Pros of Using Wind Energy	Cons of Using Wind Energy
– Include: it lessens our dependency on oil and OPEC nations – it decentralizes our energy economy – it is renewable – it is the cheapest of all renewable energy – moderately high energy yield – stable energy prices – can be built offshore – possibilities of multiple use—wind farms can also be used for agriculture – it has no carbon emission or other pollution.	– Include: energy transfer is not efficient – wind farms that are constructed in the desert may cause a disturbance to the ecosystem – they disrupt birds' migratory patterns – transmission lines attract crows that are pests to sensitive species – bats spontaneously die when they come in to proximity to wind farms – wind farms often produce more electricity than is needed at off-peak energy times, and the U.S. power grid is limited in its storage capacity – they cannot be built near military base – they can be lethal to eagles and birds of prey – cannot have wind farms near human settlement for risk of blades falling off at high speeds and causing major damage or injury – turbine syndrome—not yet recognized by the CDC (The Centers for Disease Control and Prevention)—people living near wind farms claim to experience nausea, headaches, and dizziness. It is perhaps caused by the constant hum or the physics of the turbines. More research needs to be conducted regarding this[ix] – loss of economic of value of land located near wind farms – NIMBY—most people do not want a wind farm in their back yard.

Geothermic Energy

Geothermic energy is one type of renewable energy that cannot be traced back to the sun. It makes up 0.8% of the world's renewable energy source. Deep underground in the earth's core, radioactive decay of isotopes (typically Uranium or Potassium) emits tremendous amounts of energy and heat with temperatures up to 9,000°F! At such extreme temperatures, earth's rock **(mantle)** becomes liquid **magma**. Eventually, magma flows to the earth's surface heating subterranean water and creating geysers and hot springs. This intense heat energy can be harvested as an energy and electricity source. We will discuss the process in subsequent chapters.

Places of tectonic plate boundaries are ideal for geothermic energy, or where earth's crust is thin and allows heat to penetrate. The country of Iceland lies on a fault line and experiences a relatively high amount of volcanic activity. This geography makes Iceland ideal for creating geothermic energy power plants and using geothermic energy as a primary energy source. Geothermal plants account for 25% of the electricity production in Iceland (and El Salvador for that matter),[x] and Iceland receives 50% of its energy from geothermal (for heating homes and water.)[xi] China and Turkey are the 2 countries that rely most on geothermal (up to 80% according to iea.org) In the United States, some ideal areas for harvesting heat from earth's crust include Oregon, Alaska, parts of Nevada, and California. As a result of geothermic technology becoming advanced and more suitable sites being discovered and utilized, this type of renewable energy is growing. Although the US has large geothermal resources, it is not growing as quickly as solar or wind energy.

Denis Kichatof/Shutterstock.com

But if the geography is not ideal for geothermic energy, then capturing heat from deep in the earth's crust and transferring it, even with new technology may be a tricky endeavor and the idea abandoned and replaced with other renewable resources.

Geothermic power plants are used to convert the energy in earth's crust into electricity. So how does this work exactly? Cool water is pumped into the earth's crust, picks up earth's heat through a convection process, and brings heat to the surface. Or wells are drilled into an underground reservoir that can be up to 700 degrees Fahrenheit (Source: www.eia.gov). Heat and steam cause turbines to spin, generating an electrical current that connects to the power grid.

There are even still more ways to gain access to earth's natural subterranean heat stores. More than half a million homes around the world have a **Geothermal heat pump** that enables people to use less electricity. The earth's temperature underground remains consistently at 55°F. Engineers place air or antifreeze-filled pipes underground that naturally takes on the same temperature as the ground. The air or antifreeze is eventually pumped and circulated throughout homes with the increased temperatures. Granted, 55°F is still a relatively chilly temperature and people tend to feel most comfortable at 70°F. Indoor air temperatures need to be heated up from 55°F to a comfortable temperature of about 70°F, so additional energy would be used to cover the difference (ultimately lessening reliance on electricity).

Pros of Using Geothermic Energy	Cons of Using Geothermic Energy
- Include: it lessens our dependency on oil and OPEC nations - it decentralizes our energy economy - it is renewable and not very costly - it is energy efficient - technology is developing and becoming more available - it causes little carbon emission and little pollution and pumps are available all day, every day.	- Include: energy from geothermic power plants must come from a relatively local source, otherwise it becomes expensive to transport - relatively few suitable environments exist - toxic gases like hydrogen sulfide may be released into the air, accompanied by a strong unpleasant sulfur odor, and some other air pollution - arsenic and minerals may be released into the steam - salt can build up in pipes.

Impacts of Using Geothermal, Wind, and Solar Energy on Biodiversity and Human Health

We discussed some general pros and cons of using geothermal, wind and solar energy above. Now let us get a little more in-depth in examining the specific impacts that these energy sources have on biodiversity and human health.

First of all, the expansive photovoltaic and wind farms require a large land mass, and so compete with agricultural land, protected wildlife areas, and forests. Solar farms mean less available land for wildlife, forests, and agriculture and can lead to land degradation. The disposal of wastewater from geothermal plants may contaminate groundwater and freshwater supplies affecting human and wildlife populations; and toxic chemicals that are used for the creation of solar cells can be harmful if they are handled improperly, or if they leak into the ecosystem during either use of solar cells or disposal.[xiii] These toxins released into the environment may cause various human health problems (please see chapter 8 for more information).

Water Power/Hydroelectricity

Created by the Hoover Dam, Lake Meade in Nevada supplies electricity and water supply to California. With high water demand, increasing global temperatures, and increased evaporation, Lake Meade's water level has decreased dramatically since the 1970s. As a result, the turbines have become increasingly inefficient at generating electricity. Therefore, fossil fuels have been used increasingly as an alternative energy source to hydroelectricity. Burning of fossil fuels effectuates the increase in global temperatures. This results in a positive feedback loop of less water and more fossil fuels.

Hydroelectricity is generally generated through dams which, like the aforementioned Hoover Dam, contain turbines. When water pressure becomes high, the pressure spins the turbines, generating a flow of electrons that creates electricity. Hydropower stations can come from flowing water (rivers), stored water (reservoirs), or recycled water from reservoirs. Water turbines were a 19th-century invention. This technology was seminal in advancing the efficiency of water power. By the 1920s, 40% of the world's energy came from water. Not too long after, the emergence and efficiency of fossil fuels caused hydroelectricity to lose its edge in the energy economy. Present day, 4.5% of the world's energy supply is from water, and countries such as Norway and Costa Rica use it as a major energy source. Globally, it the leading source of renewable energy. Countries that generate the most hydropower are the US, China, Brazil, Canada, Russia and India according to worldenergy.org.

Heide Hellebrand/Shutterstock.com

Lake Meade in Nevada

Dams are not the only source of hydroelectricity. The gravitational ebb, flow of the ocean's tides, and the strong resulting energy pull are also used as a means to generate electricity. And finally, steam engines were a big power source of the industrial revolution, powering factories and trains. In steam engines, water is heated to its gaseous form, steam, which enters the engine and is expanded through a turbine or piston. The "choo choo" you hear coming from a train is steam being released as exhaust.

Pros of Using Water Power	Cons of Using Water Power
– Include: it decreases our dependency on oil and OPEC nations – it decentralizes our energy economy – it is renewable, so not very costly – it has less carbon emission or other pollution than fossil fuels – high energy efficiency – dams can have multiple uses—flood control, electricity, and recreation—and have a long life span. Hydropower is the most versatile renewable energy. It is able to meet peak demand and it is consistent and predictable.	– Include: it is costly to build – it has high environmental impact – loss of land can result if reservoirs are created – the construction of dams can potentially cause masses of people to relocate from their homes – high evaporation rates (as observed in still water) causes water loss and seepage into rock beds – salinity increases with evaporation – sediment build up, causing dams to become less functional over time, and resulting in a less volume of water in reservoirs – fish spawning and other ecological processes become impacted.

Impacts of Using Water Power on Biodiversity and Human Health

There is an abundance of rotting vegetation in dam reservoirs, and this leads to the emission of greenhouse gases; the construction of dams leads to the destruction of other natural habitats such as forests and grasslands; and disruptions to river flow—dam construction can alter the availability of freshwater to humans and wildlife (depending on the dam, it can either increase or decrease).

However, the construction of dams can also inadvertently lead to the creation of wetlands which act as conservation areas for various species of waterfowl and fish.

Biomass

Biomass, at 79%, is the most abundant source of worldwide renewable energy and 11% of all energy sources worldwide. Biomass constitutes 1.5% of electrical sales in the United States, and in the United States, it made up more than 35% of renewable energy generation in 2009. Biomass energy was surpassed by wind energy as the major renewable energy source shortly thereafter.[xiv]

Biomass is considered an important component of the carbon cycle and is the remains of what was once living organisms. In the form of anything ranging from wood to corn stalk to algae to animal waste, these remains can be burned for fuel.

Why is Biomass an energy source? Through photosynthesis, the pigment chlorophyll captures sunlight to make carbohydrates—a complex organic compound containing Carbon, Hydrogen and Oxygen (see Chapter 2). These carbohydrates store a lot of potential energy in their bonds. When they are burned, they release the energy that was initially captured from the sun—and release water and carbon dioxide back into the atmosphere. If done correctly, this form of energy can grow quickly and produce little waste and use little water. It can reduce air pollution and carbon emissions that would otherwise be created through fossil fuel use.

Some examples of biomass energy include: *energy crops* that can be grown in large quantities and also are able to grow on marginal soil. Growing them on marginal soil would keep the energy crop from directly competing with agriculture for land. Many perennial grasses can be used, and sometimes can be harvested for up to ten years before being replanted. Many grasses, like switchgrass that grows in the Midwest and the South and on the Great Plains, is drought resistant, flood resistant, and can grow on poor soil. This means that it does not compete with agriculture and is ideal to grow on marginal soil. It addition, it does not require too much fertilizer.[xv] Today switchgrass is mostly used for soil cover and livestock feed, but can also be used as biopower. *Crop residues* include biomass left over from agriculture is often collected and used as biomass.

Manure—or waste from livestock—has many uses. It contains nutrients and minerals (some cultures cook with it!), and so helps maintain soil fertility by putting it back into the soil. In some cases, it can be used as an energy source through gasification, anaerobic digesters, or combustion. *Woody biomass* is byproduct of milling timber such as sawdust and bark and is often used to power mills. *Forest Debris*—such as branches, limbs, tree tops, dead trees, left over from timber milling can also be used; trees that take a short time to grow and easily rotated can be used as biofuel. And *urban waste*: waste from construction wood, tree trimmings, methane from landfills, and biodegradeable garbage are all possible candidates for bioenergy.

Switchgrass

Bioethanol is a form of energy derived from biomass. Bioethanol is made from sugarcane or other plants with high sugar/starch content. Microorganisms **ferment** by feeding on the sugar and produce bioethanol that is used as fuel and even to power vehicles in some parts of the world. Catalysts for this process include yeasts and bacteria. Biodiesel, another fuel, is biomass made from plants with high oil content such as soybeans. Biodiesel is the only biomass fuel to meet the standards of the **Clean Air Act** and to be officially approved by the EPA (Environmental Protection Agency).

Ways that Biomass is Converted to Energy

In the above paragraph, we discussed the many potential and various sources of biomass. Traditionally, biomass was used to produce steam power, but that has changed with advances in technology and now has many different uses. Now we will briefly examine an overview of new various ways that biomass is utilized for energy:

Direct combustion: This is the most common and the oldest way to use biomass to create electricity. This involves burning biomass to produce steam. Steam is used to spin the turbines which generate electricity.

Repowering: is the ways in which coal plants (and natural gas plants) are converted to run entirely on biomass.

Anaerobic digestion: In a controlled setting, microorganisms can be used to break down biomass to produce methane and carbon dioxide. Methane can be used for heat and power.

Combined Heat and Power (CHP): CHP uses the heat from the combustion of biomass to power industrial processes or heat buildings located in close proximity.

Co-Firing: This involves using biomass energy and mixing it with coal energy. This can reduce harmful emissions such as mercury and sulfur, and reduce operating costs because it only requires small adjustments to already-standing coal operating facilities.

Biomass gasification: This is when biomass is heated under controlled conditions and in the presence of Oxygen. Regulated and under pressure, it can be converted to syngas, a mixture of carbon monoxide and hydrogen. Syngas is typically refined to remove impurities and is put through a turbine to generate electricity.

Energy density: Most biomass has a large water mass content, and so is less concentrated in potential energy than fossil fuels. This means that per pound, biomass carries less energy. To make it more efficient, biomass is often dried up and made into pellets to increase energy density and efficiency.

Impacts of Using Biomass on Biodiversity and Human Health

Burning wood indoors can cause many respiratory and health issues. The agricultural demands for using biomass create the need for heavy machinery, pesticides, and herbicides. This requires the use of fossil fuels on the onset, and so all of the negative ecological impacts of using fossil fuels that we discussed would be factored in to using biomass power. Direct combustion can waste a lot of energy and create pollution. The expanding need for land to grow monocultures (such as sugarcane and fast-growing trees) takes habitats away from possible cropland and other agricultural land. Clearing land for energy cops is harmful in areas such as clearing forests, savannas, or grasslands. This is a damaging practice because it often leads to displacing food production for biofuel and compromises and depletes a carbon-rich ecosystem where carbon remains safely in the ground. This may also indirectly increase the amount of carbon in the atmosphere by taking it out of the ground; Loss of land mass and the agricultural demand of growing biomass can lead to an overall shortage of food. The agricultural demands of raising crops for biomass can contribute to chemical pollutants in the air and soil, which have adverse effects on wildlife and human health; and the burning off crop residue (slash and burn) leads to lack of water retention and nutrients in the soil.

Algaculture

Algae (collectively called plankton and a type of biomass) are photosynthetic organisms found in all aquatic habitats that form the base of marine food webs. They are also among the most diverse organisms on earth. Algae are formed as a result of a symbiotic relationship with cyanobacteria (unicellular photosynthetic microbes) that release oxygen gas and absorb carbon dioxide. Please note that when earth was in its infancy, there was very little free oxygen gas in the atmosphere. Many scientific historians credit cyanobacteria as being instrumental in creating free oxygen gas as a major atmospheric gas that it is today.

Additionally, algae can be used as a fuel source when cultivated, separated from water, and dried. In fact, there is a movement called algaculture supporting a core belief that algae could be used for jet fuel to meet our air travel needs. Consider that algae are able to produce up to 300 times more oil than soybeans.[xvi]

Nagy-Bagoly Arpad/Shutterstock.com

Wood Biomass as Fuel

Pros of Using Biomass	Cons of Using Biomass
– Include: it decreases our dependency on oil and OPEC nations – it decentralizes our energy economy – it is renewable, and not very costly – energy sources can be grown locally and with little overhead cost – energy is renewable with less pollution and carbon emissions than fossil fuels – sources can be grown in poor soil acting as groundcover to prevent runoff – there is no net increase in carbon dioxide in the atmosphere since biomass is already part of the carbon cycle – it is possible to grow some of the agriculture on marginal soil, preventing runoff.	– Include: deforestation to make room for suitable agricultural sites – air pollution from burning agriculture – low energy yield (so a lot of biomass must be used) – large water and carbon footprint created for watering biofuel crops – soil erosion from raising crops – biofuel requires a lot of land use and cannot be harvested in large quantities without displacing land used for other crops.

Hydrogen Fuel Cells

Hydrogen fuel cells are much more efficient than internal combustion engines and, possibly, a better alternative to fossil fuels, especially since they have lower and cleaner emissions. They are also conveniently small in size. Hydrogen fuel cells are based on electrochemical reactions (the flow of ions) and work in a similar fashion to batteries, since the cells have both negative and positive electrodes. They can be used to run cars, boats, mass transit, and now, possibly, aircraft.

This process uses the Hydrogen cation (H^+) for energy. Hydrogen is a ubiquitous atom found in nearly all organic molecules, so this is considered a renewable energy source. A catalyst is used to remove a hydrogen ion that is stored in water from the hydrogen atom and eventually use it to generate an electrical current which is then converted to electrical or thermal power.

The energy for the catalyst can be produced by renewable or nonrenewable energy sources. One of the obstacles of the hydrogen fuel cell is that it requires a lot of energy to remove Hydrogen, which could, theoretically, use a great deal of fossil fuel in its preparation stages, nullifying the benefits of using renewable energy.

The hydrogen must be stored either as a gas (this is potentially dangerous because it is explosive and large in volume), as a liquid (which must be stored at very low temperatures), or as crystals.

Pros: energy efficient; waste is water; clean emissions; can fuel most types of travel

Cons: It is expensive, requiring a great deal of energy in preparation stages to isolate the hydrogen cation; the technology is not widespread; more research is required to make it viable, less costly, and convenient for everybody; and storage can be complicated and dangerous.

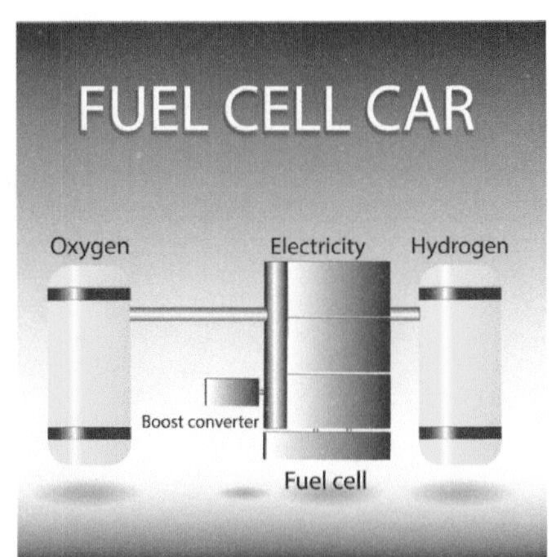

Marimaria/Shutterstock.com

Gang Liu/Shutterstock.com

Fuel Cell Concept Car

The Future of Renewable Energy

According to NREL, the United States could get 80% of its energy from renewable sources by 2050. In addition to adopting CCAs, many cities are moving toward more sustainability and carbon neutrality. Cities are moving toward renewable energy and adopting net zero energy codes for buildings allocated for residential dwelling. and incentivizing public transportation and alternative transportation with bike paths. Their buses have transitioned to natural gas sourced from landfills. And some cities like Santa Monica have adopted ordinances that solar energy is required for all new construction. What we know about renewable energy is that it is growing. Wind power tripled from 2007 to 2012.[xvii] In 2012, wind power provided almost 25% of the energy for South Dakota and Iowa, and more than 10% in seven other states.[xviii] In the United States, 29 states and the District of Columbia have all adopted renewable electricity standards, and 17 states and the District of Columbia are required to get 20% of their energy from renewable resources by the year 2025. Hawaii and Maine have the highest renewable energy standards.[xix]

With the above statistics, it appears are though we are moving toward a decentralized energy economy in which we rely less on fossil fuel and more on cleaner, renewable energy. And the National Renewable Energy Laboratory (NREL) agrees! The NREL is working closely with the U.S. Department of Energy to make renewable energy competitive with fossil fuels and other nonrenewable sources of energy.

All over the United States, photovoltaic research is conducted at universities, institutions, and laboratories in conjunction with the NREL and the U.S. Department of Energy. Their efforts are directed at decreasing capital costs of installing wind farms (which can be costly to construct) in addition to improving their energy output. They are also looking to expand hydroelectricity. Research includes innovative means to harness energy from the ocean tides, currents, and waves. Their initiatives also include working to make geothermal energy more widespread, cleaner, and affordable (nrel.gov).

In November 2015, members of the United Nations Framework Convention on Climate Change (**UNFCC**) met in Paris to discuss how to fight climate change and how to take aggressive measures to ensure an international, low carbon future. Part of this agreement includes improving and sharing technology for renewable forms of energy.

There are many other things we can do to become energy efficient and make renewable energy more widespread, like upgrade the energy grid so they are more efficient and can provide energy at peak times—and adapt to changing supply and demand. We could improve the technology to enable storage of energy for use when energy is in high demand. We could invest in the technology of batteries, thermal storage, etc. And we can look into building new transmission lines to link regional energy grids and to connect areas with lots of potential solar energy and wind energy to areas with high energy demand. The EPA is promoting a RE-powering program in which contaminated landfilled and mines are used as a site for solar and wind farms.

Fun Fact

Human ingenuity seems to know no limits when it comes to new and innovative ideas to meet our growing energy demands. Chemist Daniel Nocera, who received funding from the U.S. Department of Energy's Advanced Research Projects Agency (ARPA-E), has created an artificial silicon leaf that is able to mimic photosynthesis—using sunlight to split water into oxygen and hydrogen. This mechanism does not require access to the centralized power grid, enabling people who do not live near an energy grid to have access to electricity.

What measures can we take to move our energy economy toward a place where renewable energy sources are as widespread, cheap, and energy-efficient as nonrenewable? Instead of government subsidies for fossil fuels, the government could make renewable energy artificially cheap so that consumers have incentive to explore and use these options. Furthermore, tax breaks and concessions for renewable energy would increase incentive to shift toward a renewable energy economy. And finally, education is always a key component to becoming a responsible and informed consumer.

Review Questions

1. What is the difference between CSP and photovoltaics?

2. What are OPEC nations? Can you list a few? What are the ramifications of not relying on OPEC?

3. **Magma** and **earth's crust** are associated with what type of energy? Please explain.

4. What is the significant of the Clean Air Act?

5. List and describe the different sources of biomass/biopower.

6. What are the three parts of a wind turbine?

Discussion Questions for Class

A. How do we overcome the obstacles and move toward a renewable energy economy?

B. How could use of renewable energy improve environmental problems?

C. Which form of renewable energy seems the most viable? Which seems as though it would leave the smallest footprint?

D. What are some economic incentives of using renewable energy?

E. Can you think of any creative uses for the sun's energy?

F. New Zealand, Canada, and parts of Europe use wind energy. How does it impact their economy and quality of life?

G. What political changes might we observe, if any, by using more renewable energy sources?

H. Often steam heat from coal factories is used to reheat water. This is called cogeneration. How is this practice energy efficient? Can we use cogeneration elsewhere?

I. Are we moving toward a decentralized energy economy? Will the NREL be successful in making sure that renewable energy is competitive with nonrenewable markets?

J. Can you think of benefits of using hydrogen fuel cell as energy?

Suggestions for Your Personal Sustainability Journal

Seek out solar initiatives, resident and commercial incentives, and personal incentives that include solar water heaters, solar space heaters, and photovoltaics. Personal exemptions are available, as is reasonable and affordable financing.

Seek out energy rebates.

Consult the database of state incentives for renewables and efficiency at www.dsire.org.

In colder times of the year, set the thermostat to 68°. In warmer times, set it to 78°.

Keep drapes open during the day to maximize sunlight (in winter months.) Keep drapes closed at night to keep heat in.

Add your own suggestions here.

Endnotes

i. O Machol, Rizk. 2013. Economic value of U.S. fossil fuel electricity health impacts. Environment International 52 75–80.

ii. Solar Energy Industries Association (SEIA). 2014a. Solar energy facts: 2013 year in review. Washington, DC.

iii. GTM Research and Solar Energy Industries Association (SEIA). 2014a. U.S. solar market insight report 2013 year in review. Boston, MA, and Washington, DC

iv. GTM Research and Solar Energy Industries Association (SEIA). 2014b. U.S. solar market insight report Q1 2014. Boston, MA, and Washington, DC

v. National Renewable Energy Laboratory (NREL). 2014a. PVWatts® calculator. Golden, CO. Online at http://pvwatts.nrel.gov

vi. Global Wind Energy Council (GWEC). Global Wind Report 2012.

vii. American Wind Energy Association (AWEA). 2013. Wind energy top source for new generation in 2012; American wind power installed new record of 13,124 MW.

viii. American Wind Energy Association (AWEA). 2013. American wind power now generates over 10 percent of electricity in nine states.

ix. Abcnews.go.com

x. U.S. Energy Information Administration (EIA). 2012. International Energy Statistics. Renewables: Electricity Generation: Geothermal.

xi. National Energy Authority and Iceland Ministry of Industries and Commerce. 2006.Energy In Iceland: Historical Perspective, Present Status, Future Outlook, Second edition.

xii. Geothermal Energy Association (GEA). 2013. 2013 Annual US Geothermal Power Production and Development Report. SNL data.

xiii. Pimentel, D., Rodrigues, G., Wang, T., Abrams, R., Goldberg, K., Staecker, H., Ma, E., Brueckner, L., Trovato, L., Chow, C., Govindarajulu, U. and Boerke, S. (1994). Renewable energy: economic and environmental issues. In BioScience 44:536-54

xiv. Energy Information Administration (EIA). 2010. Annual Energy Outlook 2010. Online at http://www.eia.doe.gov/oiaf/aeo/index.html

xv. Natural Resources Conservation Service (NRCS). 2006. Switchgrass burned for power. Washington, DC: U.S. Department of Agriculture. Online at http://www.ia.nrcs.usda.gov/News/newsreleases/2006/switchgrass.html

xvi. David L. Daggett of Boeing Aerospace highlighted this potential in his 2007 report "Alternative fuels for use in commercial aircraft"

xvii. American Wind Energy Association (AWEA). 2013a. U.S. wind industry fourth quarter 2012 market report. Washington, DC.

xviii. Energy Information Administration (EIA). 2013. Electric power monthly: Data for November 2012. Washington, DC. Online at http://www.eia.gov/electricity/monthly/pdf/epm.pdf.

xix. Union of Concerned Scientists (UCS). 2011. State renewable electricity standards. Presentation. Cambridge, MA

Chapter 7

Renewable and Nonrenewable Resources

DID YOU KNOW?

The tropical rainforests are one of the earth's oldest ecosystems. In fact, many fossil records estimate them to be a few hundred million years old! As far as ecosystems go, this is considered extremely old. Take into account that North America was covered in ice until approximately 10,000 years ago, and so many of the ecosystems that exist today on the continent were created after the glaciers receded. The tropical rainforests never experienced glaciation, which helps explain their longevity.

Tropical rainforests contain over 50% of the world's biodiversity—including plants, animals, fungi, and microbes. Within this biodiversity are valuable goods that we use every day. Such goods include plants that provide us with medicinal value—antiviral, antitumor, as well as disinfectant and antiinflammatory properties; furthermore, they provide a vast variety of food from coffee to bananas. It is estimated that 80% of plants used to treat cancer are found here.

It would seem that we have great incentive to preserve these forests, but despite their immeasurable value, we cannot prevent them from being cut down faster than they regenerate. It is estimated that a rainforest area, the size of a football field, is cut down each day, and, along with it, 50 of its species become extinct. Is there anything we can do to slow down deforestation of the tropical rainforests?

Fifian Iromi/Shutterstock.com

A Tropical Rainforest

Natural resources are defined as anything that comes from the earth and is useful to humans in some capacity. Whether or not natural resources are considered renewable depends on how often we use them and how quickly they are regenerated. Air, water, some types of energy (solar, wind, etc.), plants, and animals are, for the most part, considered renewable resources because they regenerate almost as quickly as we consume them. Therefore, they are able to evade long-term depletion. Freshwater, land, and minerals are considered nonrenewable because they do not regenerate as quickly as their supply is used by humans, and are at risk of diminishing returns, if not, complete depletion.

Forests

Forests have both economic and ecological value that is difficult to quantify. Economically, they provide crops, timber, medicine, and eco-tourism, among many other things. Ecologically, they moderate climate, are instrumental in chemical cycling, provide habitats for both plant and wildlife, and act as a major carbon sink.

An Old-Growth Forest

Rainforests, as well as other forests (deciduous, old-growth, coniferous) provide timber that is used to build homes, ships, furniture, paper, etc. Most forests—especially **old-growth forests** that have not been touched or disturbed by humans in at least 100 years—are nonrenewable resources. Old-growth forests (sometimes called primary growth forests) are characterized by having little human activity, older trees, indicator species, rich and fertile soil with a healthy fungal ecosystem, and lots of dead, decaying biomass. They also often have epiphytes such as moss growing on them and contain an abundance of biodiversity.

In the last 8,000 years, 46% of the world's forests have been cut down, most of this occurring since the 1950s in tropical climates. The tropical rainforests with their tall stature and large canopies provide us with timber and crops like coffee, tea, nuts, and fruit as well as the plants with pharmaceutical properties mentioned above. Besides the unfortunate direct loss of wildlife and biodiversity, another side effect of deforestation is the **edge effect**. This occurs when, as a result of deforestation, a larger proportion of land

The Effects of Clear Cutting

receives excessive sunlight. Edge effect causes **desertification** of the land, when once-fertile semi-arid lands turn infertile and unable to grow crops or vegetation. Desertification makes it difficult for plants and wildlife to grow and flourish. It has also been linked to reduced rainfall in the tropics, as transpiration rates dwindle. Could deforestation lead to the tropical rainforests drying up?

One method in which trees are harvested is **clear cutting**, when a large proportion of trees are cut down in one fell swoop and the ground is left exposed. Clear cutting is the most common type of deforestation and perhaps the most controversial. The pros of this practice include generating a high timber yield (and therefore profit) in a short amount of time. It also enables a fast-growing replacement tree farm to be grown in its place. Unfortunately, tree farms do not compensate for the biodiversity lost during deforestation, but they can be useful as a means to control runoff and provide future timber. As a con, clear cutting greatly reduces biodiversity and also causes major runoff that pollutes lakes and streams. Furthermore, after the trees are cut down, they are often burned, releasing the greenhouse gases, carbon dioxide and carbon monoxide, into the atmosphere. Another method is **strip cutting** in which a corridor is cut through a patch of forest land, usually upslope from a body of water (so intact land absorbs runoff and prevent water pollution). This prevents massive runoff from occurring but compromises biodiversity and disrupts species' ranges. And finally, **selective cutting** is the process in which only the most desirable trees are targeted and cut down. Selective cutting is considered the most sustainable tree harvesting method, as it causes the least amount of disturbance to an intact ecosystem. But it is not without its own controversy. The selected trees are often the oldest, and home to many species.

There are many adverse effects to logging. After deforestation, forests are unable to offer their ecological services—since there is no biomass to retain water flooding may occur, contributing to runoff and desertification of the soil. Forestry and logging account for 17% of Global Greenhouse Gas emissions.[i] Furthermore, the access roads that are built to facilitate logging act as conduits for other activities and factors that are harmful to the ecosystem and indigenous people who live there. The ready-built-roads often bring illegal loggers, illegal hunters, and bring in disease and pathogens to people and animals that they have never been exposed to before. This means that they have not developed and do not have the proper immunity.

Christopher Kolaczan/Shutterstock.com

Strip Cutting

Tree farms are an alternative to deforestation. This is the deliberate planting of trees with uniform size and age for commercial use. Usually, the tree of choice is fast growing, such as pine or eucalyptus. Although they do not have as much biodiversity as old-growth forests, tree farms can be beneficial to the environment because they cover the soil, preventing runoff and desertification, and they provide commercial use of wood as well as wildlife habitats.

A Eucalyptus Tree Farm

Minerals

As we will examine in subsequent paragraphs, earth is a very dynamic, volatile structure made of three concentric layers. The innermost layer is the earth's **core**, comprised of a hot mass of liquid metal called magma. **Magma** mostly consists of the heavy metals nickel and iron. The middle and the thickest layer of earth (roughly 75% of earth's volume) is called the **mantle**. The earth's mantle is made of lighter elements than are found in the core: some of these elements include magnesium, oxygen, and silicon. The mantle is composed of semiliquid rock of very hot temperatures, yet cooler than the earth's core. We might compare the mantle's consistency to hot tar that is used to pave a driveway—before it has cooled and solidified. The outermost layer of earth is the earth's **crust**, constituting only 1% of the earth's volume. You may compare the earth's crust to the skin surrounding the mass of a fruit. The earth's crust is a cooler temperature than the more inner layers. It is constantly being recycled through wind, water, erosion, and attrition.

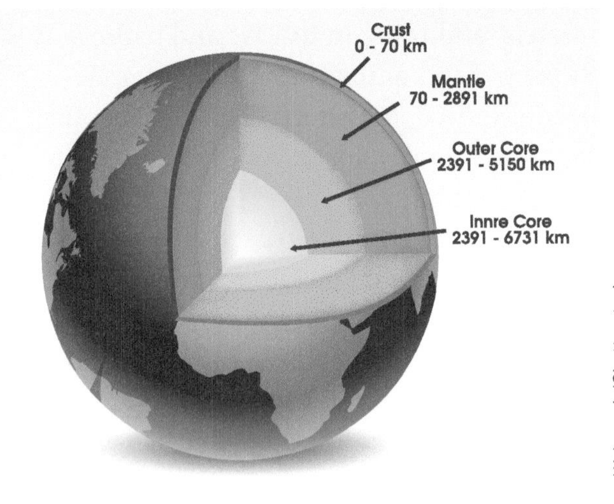

Within earth's crust, we find three major types of rock: sedimentary, igneous, and metamorphic. (Rock is comprised of more than one mineral, which we will discuss below.) Over time, the dust, sediment, and debris erode from earth's surface as a result of wind, water, and attrition. Eventually, this sediment settles, and over thousands and millions of years, is subject to heat and pressure. Eventually, the sediment mineralizes to create **sedimentary rock**. Sedimentary rock, as we touched upon in Chapter 3, provides us with the fossil record which was instrumental in shaping the theory of evolution. Examples of sedimentary rock include limestone, sandstone, and conglomerate.

PRILL/Shutterstock.com

Limestone

Igneous rock is hardened magma that has leaked from earth's core, often through volcanic activity. Examples include granite, obsidian, pumice, and basalt.

And finally, **metamorphic rock** was once either sedimentary or igneous rock. Through time, with heat and intense pressure, such rock morphed into a new form. Marble, for instance, is transformed limestone.

Billions of years ago, earth's mineral deposits were formed as a result of hot magma erupting in the ocean floor. The chemical cycles were much different than they are now, but ended up as the mineral deposits accessible to us present day. Minerals are inorganic substances (not carbon-based) and, with exceptions, are considered nonrenewable resources. We need minerals not only as dietary supplements (zinc and potassium) and precious gems (sapphires, rubies, diamond) but for everyday needs such as glass, pots, pans, and electricity.

Siim Sepp/Shutterstock.com

Obsidian

Wjarek/Shutterstock.com

Steps in a Marble Quarry

Minerals are a result of earth's ancient geological history, can be extremely very valuable, and are found in finite supply in the earth's crust. Minerals are found in metal form, such as copper and aluminum (metals conduct electricity), but minerals can also be nonmetals, such as salt and sand (sand is used to make glass). Minerals take millions of years to form, so when their supply runs out as a direct result of overconsumption by humans, their reserves become completely depleted. Thus, minerals are considered a nonrenewable resource.

Copper is a mineral used in homes for electrical wiring, house fixtures, pots, pans, and many other uses. Copper is found in the earth's crust, often with other metal deposits such as lead and gold. Copper has been used for thousands of years, but almost 95% of copper was mined in the last 100 years. Once copper reserves are depleted, they are gone forever.

Rock is a substance comprised of more than one mineral. **Ore** is a type of mineral in which metal is found in high concentrations. We mine for ore because it is useful to our everyday needs. Although ore is only a fraction of the earth's surface, it is very important economically. Please refer to Chapter 5 for the different types of mining and its effects to the environment.

As mineral stores become depleted, a lower-grade ore is mined, requiring more processing and creating more pollution. Minerals such as aluminum, copper, iron, calcium, and magnesium are found in the earth's crust. But minerals such as zinc, copper, silver, gold, and tin can also be found in small concentrations in the oceans. They are the result of attrition of rock whose debris has been deposited into the ocean. Some good news is that most metals are recyclable, and so their depletion times vary or can be prolonged.

Fun Fact

Minerals are not only economically important but also offer great insight into earth's history and the evolution of many species. The fossil record, for example, was instrumental in helping Charles Darwin form his theory of evolution through natural selection. Similarly, earth's geological record can give us insight into earth's early atmosphere; the biggest iron deposits in the world are found in South Africa and Australia—in the form of iron oxides—or what we call rust. Scientists are studying these iron oxide deposits to make predictions about the composition of earth's early atmosphere and the bacteria that helped form them.[ii]

A Dynamic Earth

The earth is ~4.6 billion years old and is constantly morphing and undergoing profound changes to its surface. In such an incomprehensibly vast amount of time it is hard to imagine the changes that may have occurred and altered the face of earth. Mountain ranges and islands have formed and disappeared, continents have shifted, and ocean currents have altered course, not to mention the constant and cataclysmic volcanic activity.

These phenomena affect both the living components of earth (the **biosphere)** and the nonliving geologic components. The earth's outer rocky solid surface, called the **lithosphere**, is constantly being recycled—created and destroyed through both internal and external powerful forces. Some such *external* forces include the creation and recession of glaciers, rain, wind, and erosion. Erosion causes the attrition of rock and deposits minerals in the soil.

Because Earth's core is made up of hot liquid rock, some *internal* forces help shape earth's surface, including heat and pressure that build up from earth's interior. The heat and pressure build and intensify, applying pressure to its surface and causing major shifts and changes in earth's geological cycles.

The continental and oceanic plates are less dense than the **asthenosphere**, the warm, less dense layer found below plates. Because of this, plates are subject to change and movement. The slow movement of the outer layer of the earth is called **plate tectonics.**

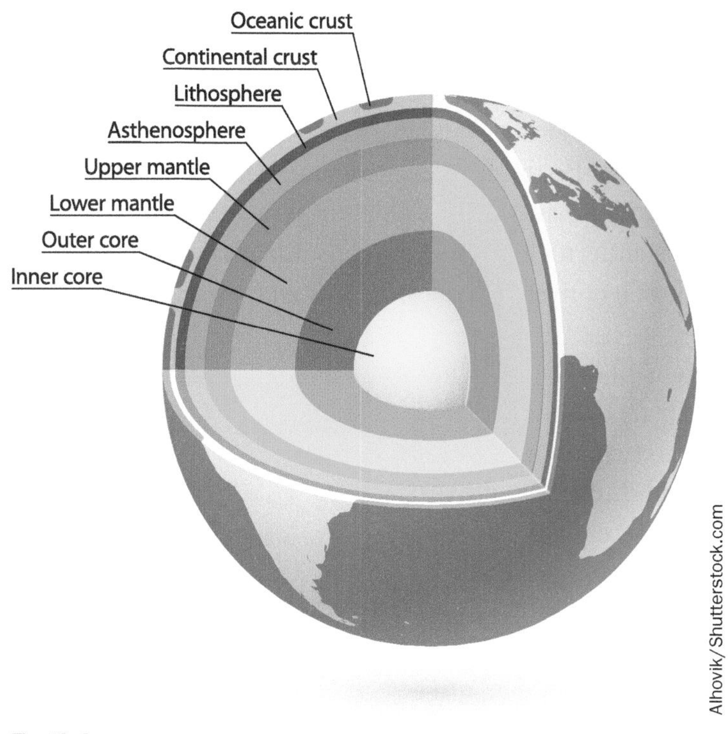

Earth Layers

Plate tectonics has had tremendous effects on the earth's surface and its individual ecosystems. As plates move, continents move apart, altering species' ranges and habitats and altering weather patterns. It also creates volcanic islands that are hotspots and home to many endemic species.

Consider the information presented about the supercontinent Pangaea. At one time, South America and North America were connected to Africa, but over time, plates shifted and the continents separated. This shift also affected the circulation of ocean currents, which carry heat and affect temperature and climate. It also had profound effects on atmospheric cycles, which, in turn, have major effects on climate.

In terms of the earth's crust, there are two distinct types of crust: the oceanic crust and the continental crust. The oceanic crust, made of mostly basalt, lies beneath the ocean and is thinner than the continental crust. The continental crust, made mostly of granite, lies beneath the continents and is thicker than the oceanic crust.

The Various Plate Boundaries

Divergent boundary occurs beneath the ocean when two plates move apart at the divergent plate boundary, lifting the two plates. At the upswelling where the two plates diverge, there is a crack or a fissure where magma flows out. Eventually, magma cools and the process is repeated. Most ocean ridges are a result of this, and the mid-Atlantic ridge is an example of the result of a divergent boundary.

A **convergent boundary** between an oceanic plate and a continental plate occurs when the two plates collide from internal forces pushing both of them together. In this case, the thicker (oceanic) plate slides under the thinner (continental) plate. We call this a **subduction**—the plate is forced down into the mantle where the temperatures are extremely high, melting the plate on the bottom. Mantle melts and eventually magma reaches the surface. The Andes Mountains are a result of a convergent boundary as well as the Cascade Mountains of North America.

When a convergent boundary occurs between two oceanic plates, the two plates come together and one (usually the older and thinner plate) **subducts**, or moves under, the other. Magma reaches the surface and cools forming islands. The Eastern Caribbean Islands formed as a result of two oceanic plates converging.

An **oceanic trench** occurs when two oceanic plates collide and one plate begins moving below the other, causing deep depressions and fissures in the ocean floor. We find a great deal of extreme and curious aquatic life in oceanic trenches.

A **convergent boundary** between two continental plates is powerful and does not include a subduction. Both plates are forced upward, often creating mountain ranges. The Himalayas and Appalachians are an example.

THREE TYPES OF PLATE BOUNDARY

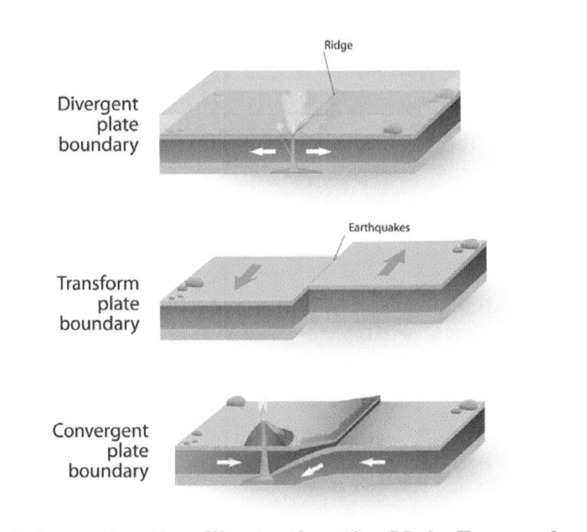

Designua/Shutterstock.com

A Cross Section Illustrating the Main Types of Tectonic Plate Boundaries: Convergent, Divergent, or Transform. Mountain building and Volcanism occur between tectonic plates at convergent boundaries with a collision or a subduction.

A **transform fault** is an area of high earthquake activity resulting from plates grinding past each other, creating enormous pressure and ultimately causing the two plates to move away from each other at the boundary. The lithosphere will absorb the energy and stress until the pressure becomes too great. Energy release causes a vibration that moves through the lithosphere in all direction called a *seismic wave*. The San Andreas Fault in California is an example of a fault line.

Land and Soil

Land

kenkister/Shutterstock.com

Land is considered a nonrenewable resource. When land it is no longer of use because its soil no longer contains nutrients, it takes many years to restore its function and fertility. This phenomenon is called desertification—when soil can no longer hold water and the soil contains little if any nutrients. This is often caused by overgrazing, industrialized farming, irrigation, and deforestation.

About 28% of land in the United States is owned by the Federal Government. Land is of ecological, environmental, recreational, and economic importance. The Federal Government uses the land for research, conservation, wildlife protection, logging, mining, and, of course, recreation. It is also used for oil and natural gas extraction.

Worldwide, there are roughly 85 people per square mile of land, and 94% of the world's population lives within 60 miles of the ocean, to the detriment of coastal areas. This leads to coastal overdevelopment and the paving over of storm buffer zones, which, in turn, leads to ineffective buffer zones and flood control, habitat loss for marine life, and loss of spawning areas for fish.

Soil

The **pedosphere** is the entire earth's sphere that encompasses soil. The earth's soil is something that we often take for granted. Perhaps we dismiss it as pesky dirt under our feet, or mud that we track into our homes. But soil is so important for sustaining life, that poor soil quality is linked to famine, hunger, and suffering.

Soil has innumerable functions that benefit all life: it facilitates growth of our crops, trees, and vegetation; it acts as a major carbon sink; it is an integral part of the biogeochemical cycling process; and it acts as a sink for many nutrients; it facilitates gas exchange; it acts as a storehouse for our waste material (landfills); and it is estimated that 95% of our food comes either directly or indirectly from soil.[iii] Soil harbors 1/4 of the world's biodiversity if it is healthy. Healthy soil may contain nematodes, vertebrate, earthworms, insects, bacteria, mites and fungi. Soil is far more complex than we might give it credit for.

Soil also has many medicinal values! In fact, many species of animals—elephants, birds, primates, and even people—are known to eat soil to soothe gastrointestinal discomfort or malaise. Some of the minerals found in soil have antidiuretic or antibloating properties and help relieve constipation. Species also eat soil because clay is known to keep toxic substances, often secondary compounds in plants, from being absorbed by the gastrointestinal tract.[iv,v] The process of eating soil is known as **geophagy**.

Furthermore, please consider these facts: There are 15 soil nutrients required for crop growth. A shortage of any one can compromise crop yield. It can take 1000 years to create 1 centimeter of soil. By the year 2050, agriculture yield must increase by 60% globally. If soil was managed more sustainably, it could yield 58% more food.[vi]

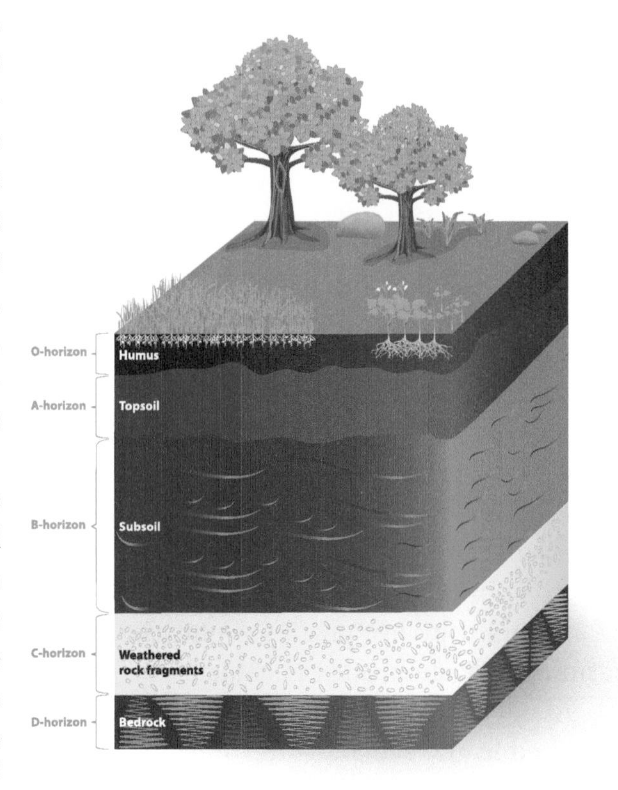

Soil Horizons

The pedosphere is dynamic, variable, and difficult to classify. Soil varies in its composition but it is usually characterized by the composition and thickness of its upper layers. Water (as groundwater) and air are found in pores between solids in its lower layers.

Soil is collectively made up of clay, silt, sand, gravel, water, air, and dead organic material from flora and fauna. It is divided horizontally into several layers, or what we call *horizons*. As soil layers progress from earth's surface deep into the earth, soil composition slowly shifts from organic matter to fine minerals to bedrock. Below is a brief summary of the soil horizons, but it is important to keep in mind that composition varies greatly and is subject to change:

Horizon O—named for being rich with humus or dead organic matter (such as decomposing leaves and twigs and plant and animal parts) and living organisms (such as bacteria, earthworms, fungi, protozoa, and insects). Wetlands tend to be rich in organic matter. Manure and plowing replenish the organic matter.

Just below this layer, we find *Horizon A*, which is considered **topsoil**. *Typical* topsoil is made up of air, water, minerals, and an abundance of organic material and microorganisms.

Horizon B is mostly minerals, but also consists of some humus and organic matter. It is rich in clay and is also comprised of silica, iron, and carbonate. Soil here tends to stick together because of the presence of clay, which is moist and sticky. Clay has a net negative charge so plants can perform **cation exchange**—the process in which plants obtain nutrients. Cations are released from plant roots while they absorb nutrients from the soil. The sticky, viscous nature of clay holds nutrients in proximity and facilitates the exchange. Some of these cation nutrients include potassium, calcium, and magnesium.

The **rhizosphere** is a complex layer of the soil that comes in contact with plants where nutrient exchange occurs. This typically occurs in Horizon A and some of Horizon B. Here in the rhizosphere, one finds mychorrhizae which assists plants in nutrient uptake and has been instrumental in helping plants colonize land.

Further down, *Horizon C* consists of minerals derived mostly from broken down rock.

The final and deepest *Horizon D* is solid bedrock.

Is soil a renewable or nonrenewable resource? This is not an easy question to answer. As mentioned above, the topsoil is the part of the soil that holds nutrients and is necessary to sustain life. Topsoil can take a few years or hundreds of years to regenerate. When semi-arid soil and land can no longer hold nutrients, we say that the soil has **desertified**. Human life and well-being hinge on the quality of the soil, so desertification is not something that can be gambled on or taken for granted.

Erosion is mobility of soil, usually through wind and water, which ultimately strips it of its nutrient content. Erosion is caused by many factors, including weather phenomena (such as wind and rainfall), over-grazing, and human activity (including urban development, slash and burn agriculture, and other such disturbances). This leads to desertification of the soil. Desertification is happening at an alarming rate, especially in urban areas or areas used for grazing of livestock.

The consistent rotation of grazing animals is an effective means to bring soil back to life. Grazers' activity, dunging and kicking up and moving around the soil, effectively fertilizes and plows the soil (as long as they are rotating and do not overgraze!)

Other ways to maintain soil nutrient content soil is to have virtually no tillage, grow organic, engage in agroforestry, and engage in more sustainable agricultural processes. For more information and further reading, please see Chapter 11.

Dirk Ercken/Shutterstock.com

Soil Erosion due to Over-Grazing and Ultimately Leading to Desertification.

Water

About 71% of the earth's surface is covered by water of which 97% is found in the oceans, 2% as glaciers or icecaps, 1% as freshwater, and an even smaller percent available to us as drinking water.

The water, or hydrologic cycle, is powered by solar energy. Turnover of water molecules in the atmosphere is approximately 9 days, but it is estimated that water molecule turnover in oceans is 3,000 years. Because turnover of the hydrologic cycle is relatively quick and water supply on earth is abundant, water is considered a renewable resource. However, because less than 1% of water is available for consumption, drinking water is considered a nonrenewable resource.

There are three Oceans; the Pacific (both the largest and the deepest with its underwater trenches) the Indian Ocean, and the Atlantic Ocean. Oceans provide use with food and recreation, act as the biggest carbon sink, and also help influence earth's climate. Prevailing winds over the oceans moderate climate as **gyres** bring warm water from the tropics to the poles. We will examine water in more detail in Chapter 9.

djgis/Shutterstock.com

Volodymyr Goinyk/Shutterstock.com

I_V_Y/Shutterstock.com

Food

We obtain our food from fishing, raising livestock, and growing crops. Due to its ample supply, food is considered a renewable resource—but the means in which we produce it in some parts of the world is unsustainable.

Taiga/Shutterstock.com

There are over 7 billion people on the planet, and we produce more calories per day today than we produced in the mid-1900s. And the population is expected to grow to 9 billion by the year 2050. But despite this, people are still not getting enough to eat due to poor distribution. Not getting enough to eat and therefore lacking nutrients is called **food insecurity**.

According to FAO, 821 million people on earth (that is roughly 10%) were undernourished in 2017. All regions in Africa, parts of Latin America and parts of Asia are hit the hardest are the most impacted by hunger and fared worst in this statistic. But most developing countries—especially in Southeast Asia and Latin America—have been successful in actualizing Millennium Development Goal 1 (MDG1) of eradicating worldwide hunger. (Please see Chapter 15 for more information.)[vii]

Macronutrients are substances that the body needs a large supply of, such as essential fats, proteins, and carbohydrates. **Micronutrients** are substances that the body needs in small amounts, such as iodine, iron, vitamins, and other minerals. A lack of both macro- and micronutrients leads to malnutrition among other health problems (an iodine deficiency leads to goiter, a lack of vitamin D and dairy leads to rickets, and a lack of vitamin C leads to scurvy, for example). It is estimated that 2 billion people globally are deficient of important vitamins and minerals in their diet; particularly Vitamin A, iron, zinc, and iodine. This is attributed to a lack of micronutrient-rich foods such as fresh fruits and vegetables.

The more developed countries (MDCs) are big meat-eaters that consume plenty of protein for consumption with the vast number of animals we raise; but raising livestock creates pollution and uses a lot of water. (The least-developed countries (LDCs) typically rely more on fish rather than on meat-based diet.) The consumption of a meat-based diet uses twice as much water as a plant-based diet according to UNESCO Institute of Water Education. And according to International Water Management Institute, meat-eaters use roughly 5000 liters of water a day while people consuming a vegetarian diet use 1–2000 liters. And consider that according to the Food and Agriculture Organization ofthe United Nations (FAO), in 2014, it was estimated that we produced 312 million tons of meat, with 68 million tons of that being bovine meat![viii] And it does not end there. Much of our fossil fuel is used to raise livestock, and a major negative impact of raising cattle is increased concentrations of methane, which is a greenhouse gas. Livestock releases 80 million metric tons of methane per year, 28% of global methane emissions.[ix] And keep in mind that methane is 20 times more potent as a greenhouse gas than carbon dioxide in a 100 year period and 84 times more potent than carbon dioxide in a 20-year period. Some sources, such as the think tank 'WorldWatch' rank animal agriculture much higher in terms of greenhouse gas emissions.[x] Their comprehensive research included examining the release of methane, nitrous oxide, and water vapor. Their research concludes that animal agriculture is responsible for nearly half of all global greenhouse gases! The methane released by cows is not only a greenhouse gas; it contaminates, and is a major polluter of soil and groundwater. Animal agriculture is also the major driver for deforestation in the Amazon Rainforest, which acts as an invaluable carbon sink and also provides a major source of Oxygen. Furthermore, up to 56% of the world's freshwater supply goes to feed livestock in the

Freer/Shutterstock.com

Source: http://www.worldwatch.org/files/pdf/Livestock%20and%20Climate%20Change.pdf

United States. Is this an efficient system, given that 1 billion people are without access to clean drinking water, and over 800 million suffer from food insecurity?

Most crops in the United States are cultivated through **monoculture**, the harvesting of a uniform type of crop. This requires lots of irrigation, and our freshwater supply is limited. Consider that it takes 1,800 gallons of water to produce 6.5 pounds of grain. Industrial farming increases the competition for land and water use, which is unsustainable given the growing population.

Monocultures also contribute to large numbers of a specific pest. This is usually counteracted with pesticides. Fertilizers, pesticides, and herbicides all use phosphorus and nitrogen. These chemicals leach into the soil and eventually harm aquatic life and water quality. Pesticides and herbicides have chemicals that also harm beneficial organisms on land and limit terrestrial biodiversity.

Organic farming methods use natural herbicides and pesticides and include **crop rotation** to conserve soil fertility. Organic farming methods also exclude genetic engineering, or the production of **transgenic crops**—crops with DNA from two sources. Crops are usually engineered to be resistant to pests and herbicides, tolerant to drought and salt, as well as the addition of vitamins. (We will discuss genetic engineering later in Chapter 12.)

Case Study

To cork or not to cork? That is the question.

A common hot-button debate among many wine connoisseurs is whether to utilize cork *or* modern screw caps—with wine and champagne bottles. The cork controversy has many tiers and environmentalists support both sides of the argument. Cork is technically considered a renewable (and biodegradable) resource if, and only if, it is harvested sustainably. Cork, which matures every 10–20 years, is taken from the outer bark layer of the Mediterranean cork oak (*Quercus suber*). It does not harm or kill the tree if done properly. And in fact these evergreen **angiosperms** can live up to 300 years, even when harvested.

LianeM/Shutterstock.com

Cork Oak

Although traditionally associated with wine, cork has many other uses such as providing material for furniture and flooring. Widespread cork use, many argue, should be encouraged as it facilitates the conservation of cork oak forests—rich with biodiversity. These oak forests are home to many endangered, sensitive and endemic species including birds such as eagles and mammals such as the lynx.

In addition to its intrinsic and ecological value, such biodiversity also provides economic value. These forest resources are a major component for stabilizing the economy of countries like Portugal and Spain where cork oak groves are found. Using cork oak groves provides jobs to thousands in the region. Those employed for forest-related jobs naturally become stewards of the forest and have incentive to preserve them and to prevent erosion and desertification of the soil.

Cork harvesting gets support from the WWF (World Wildlife Fund established in 1961), a conservation and international fundraising group whose mission is to reduce biodiversity loss worldwide. According to the WWF, "Cork oak landscapes are one of the best examples of balanced conservation and development anywhere in the world. They also play a key role in ecological processes such as water retention, soil conservation, and carbon storage."

inacio pires/Shutterstock.com

Harvested Cork Oaks in Portugal

Unfortunately, many harvesting practices result in carelessly cut into the vascular layer of the tree—the inner layers that transport nutrients. Ultimately it ends up compromising and killing the oak. Dead cork oaks are of no value to anybody.

Wine production companies often prefer screw caps and to avoid cork taint—parasite that lives in cork oak bark and ruins wine—but is of no risk to human health.

What are your opinions regarding the ongoing cork debate? Are you in favor of using cork or not using cork tops in wine bottles?

Fisheries

Fifteen percent of the world, mostly comprised of least-developed countries (LDCs), relies on fish as a major dietary protein source. With heart disease, obesity, Type 2 diabetes, and other health-related ailments on the rise, many health professionals and government agencies are recommending a shift from terrestrial meat toward a dietary increase in seafood consumption. This huge demand for fish has escalated to the point at which 70% of the world's fish used as a food source is facing imminent extinction.

Andreas Altenburger/Shutterstock.com

Bycatch

There is no international agency governing the oceans, so, unfortunately the most opportunistic fishing methods often win out, preventing sustainable fishing practices. Currently, trawlers are commonly used to catch fish. Trawlers are large nets that drag across the ocean floor and displace many of the species found there—such as mussels and crabs—and damage sea beds. Inadvertently, sharks, rays, and juveniles of many fish species get caught in these nets as **bycatch** and perish. This not only causes a loss in biodiversity but also disrupts the marine food web, thereby threatening dolphins, sea turtles, and sea birds, among many more. Furthermore, derelict fishing gear can get lost in the oceans and continue to "fish," even though there is nobody present to retrieve nets. Many marine animals accidently become entangled and never make it out of the nets. (Please the case study on Ghost Fishing below.)

The Marine Stewardship Council (**MSC**) is a nonprofit, international agency that works with scientists and fisheries to regulate and set standards to ensure that fishing practices are carried out sustainably. They also ensure that biodiversity and a healthy number of fish are maintained in our oceans, making sure these aquatic environments are ecologically healthy. They reward and promote fisheries that are certified. To be considered certified, fisheries must reduce the number of trawlers, reduce the bycatch of birds and fish, and educate and raise awareness of the impacts overfishing has on the ecosystem.

The problem with sustainable fishing is that it mostly focuses on and target—or the "catch." It also does not take into account the amount of bycatch—the countless animals caught in nets and left to die from lost nets (ghost nets). Since marine animals typically decompose after 3–4 weeks, it is almost impossible to predict the number of animals that are continually caught in nets lost at sea.

An alternative to overfishing is fish farming, or **aquaculture**, which operates more like agriculture than fishing practice. Developing countries rely more on aquaculture than developed nations. According to a report by the International Food Policy Research Institute and World Bank, aquaculture is predicted to provide 75% of the world's fish supply (as of 2009, they provided less than half).[iii] Aquaculture reduces the need to import fish and, if done properly, can decelerate the worldwide overfishing problem. It provides jobs and improves the economies in coastal areas. In addition, health professionals believe that there could be an overall improvement in health if more people shifted toward a seafood-based diet rather than land animal meat (like beef).

A notable success of aquaculture is the improvement of the Potomac River through oyster aquaculture. The oyster reefs were restored to make room for the oyster farms. The oysters restored overall health by cleaning out the algae from nitrogen pollution.[iv]

Unfortunately, the way that fish farms are currently run causes a lot of ecological damage. There is a much higher concentration of fish population inhabiting these farms than one would normally find in the wild. Farmed fish contain the same contaminants and heavy metals found in wild-caught fish. This artificial overpopulation not only leads to disease and pollution that spread through the aquaculture but also may spread to other natural marine habitats. Furthermore, aquaculture can be ecologically problematic if the farmed fish get out and hybridize with fish in the wild, contaminating their genetic pool. Many farmed fish are not endemic to the area in which they are farmed. The farmed fish are given antibiotics that can not only lead to antibiotic resistance but also spread to the water supply. This could potentially leak out and cause wild fish populations to be exposed, compromising wild populations and contaminating their gene pool.

Vladislav Gajic/Shutterstock.com

Fish Farms

Perhaps we need more natural and integrative measures to help solve the problem of overall health in fish farms. As we mentioned in Chapter 4, wrasse fish swim in and out of the mouths of moray eels in a mutually symbiotic cleaning ritual. It has been recently noted that wrasse fishes are also effective at cleaning out ectoparasites from the mouths of farmed salmon, mutually benefitting both species. Perhaps this mutualistic symbiosis will be a more effective at maintaining health levels and salmon, and keep harmful chemicals out of fish farms. (This is yet another example of why it is important to understand and grasp the interconnectedness and complexity of the natural world, so we can mimic its natural structure to the best of our abilities and thereby reduce our use of synthetic and harmful chemicals.)

Shrimp farms tend to be particularly destructive in other ecological ways to sensitive areas with high biodiversity. They are farmed in mangroves, which are wetlands that harbor much biodiversity and are among the most productive ecosystems. The shrimp farms damage mangroves by polluting and decreasing their natural biodiversity.

Another obstacle to aquaculture is that some fish will not breed in captivity. There are too many variables that go into fish spawning to determine a single variable that is *the* single obstacle to successful breeding, so it is a difficult obstacle to remedy. Also, fish imported to the United States from other countries may be raised under different standards and may use regulatory drugs or chemicals not allowed by the USDA (United States Department of Agriculture).

Furthermore, the productivity of fish farms that cultivate animals with short life spans (such as sardines and shrimp) may be negatively impacted by climate change.[v] Shrimp and sardines eat phytoplankton and algae, small photosynthetic organisms. Anything photosynthetic contains chloroplasts or plastids, subcellular structures that produce chlorophyll (the pigment involved in photosynthesis). Chlorophyll production and photosynthetic rates are greatly disturbed and often erratic and reduced by change in temperatures. This implies a drop in food source for the farmed fish. It also demonstrates the interconnectedness of marine ecosystems and food chain, and how climate change profoundly affects them.

Case Study

GHOST FISHING

Are Ghosts Haunting Our Oceans?

What happens when fishing gear gets lost at sea due to the unexpected wreck of fishing boats or the accidental abandonment of fishing gear? Do nets keep fishing although there is nobody present to man them? It would appear that the sad answer to that is "yes."

Ghost Fishing is a nonprofit organization comprised of volunteer divers and other committed topside support staff. Together, they retrieve lost and abandoned commercial fishing gear at sea. Divers often retrieve thousands of pounds of nets, requiring a good deal of man power to haul up on to the deck. The nets are recycled into various products such as carpet flooring, socks, and bathing suits.[vi]

This agency is aptly called *Ghost Fishing* because these large nets continue to "fish," entangling aquatic life that get caught in them—even though there is nobody there to man or retrieve them. Sometimes this continues over the span of hundreds of years. Sea lions, lobsters, whales, sea cucumbers, octopi, eel, crab, and sea bass are examples of species that have been found caught in nets over the years.[vii]

Courtesy of Tom Boyd

Fishing Indiscriminately; Ghost Nets Kill Innocent Marine Life such as this Sea Lion

Some of the aquatic life are found alive and set free. In fact, volunteers save thousands of lives by physically untangling the aquatic life from nets. Sadly, though, other animals come to an untimely death due to their inability to escape. Sea lion bones and remnants of other sea creatures are often found caught in the derelict net.

Saving wildlife can be a dangerous endeavor since the nets are not only a danger to the sea life, but to divers as well. Divers have to be extra careful while maneuvering these thousand-pound nets, lest they become the next casualty of ghost fishing.

In addition to direct-action conservation, *Ghost Fishing* has an outreach program that involves speaking at schools and getting involved in communities to spread awareness. It has published articles in various journals and has been featured in various media outlets. Recent examples include MSNBC's "The Cycle" and features on the Discovery Channel!

To help make more change, *Ghost Fishing* is currently working with lawmakers to help write legislature and change some of the existing laws to enforce that lost commercial fishing gear is reported sooner than later. This could save many more marine lives in the future by ensuring swift and immediate action. For more information, check out www.ghostfishing.org.

"The incessant demand for seafood is what fuels the problem. If people want to make a difference, they don't need to volunteer—they just have to stop eating sea creatures."—Heather Hamza of Ghost Fishing

Review Questions

1. What is the difference between the rhizosphere and the pedosphere?

2. What are the different soil layers from the uppermost layer downward?

3. What are the different layers of earth from the inside/core to the outer-most layer? What is the composition of each layer?

4. What is a micronutrient? What is a macronutrient? Provide examples of each. Define 'food insecurity.'

5. Define transform fault, divergent fault, convergent fault, and oceanic trench.

6. Define "geophagy." How does it help the livelihood of animals?

Discussion Questions for Class

A. How does strip cutting and clear cutting disrupt species' ranges and increase the edge effect?

B. How has our overconsumption of earth's resources contributed to environmental problems?

C. 90% of the human population lives within 100 miles of the coast. How does this affect our coastline?

D. Cows give off methane, a greenhouse gas. How does an overconsumption of meat by humans affect the earth?

E. What are the ramifications of the amount of carbon and water needed to raise 1 pound of beef?

F. Hydroponics is growing plants or crops in a soil-less medium. How might this practice help the environment, if at all?

G. What is the significance of the MSC?

Suggestions for Your Personal Sustainability Journal

Shift to a plant-based diet!

If you plan on getting engaged or married in the near future, consider a vintage engagement ring (or vintage jewelry for any occasion). This cuts down on pollution and disturbance associated with mining for diamonds. It also avoids the unethical practices associated with the diamond business today.

Use composting to recycle vegetable waste.

Buy organic fruits, vegetables, and meat when possible.

Do research to discover natural herbicides and pesticides (such as marigolds, lavender oil) instead of using synthetic chemicals.

Buy recycled paper.

Buy clothes that can be washed rather than dry-clean only.

Look into ways to recycle copper, especially old copper wires.

Make sure your home has efficient insulation and that appliances used for heating and/or cooling are cleaned and working optimally. Change filters when necessary.

Almost 1 billion people on earth do not have access to clean drinking water. Volunteer and donate to charities that specialize in providing clean drinking water in the developing nations.

Add your own suggestions here.

Endnotes

i. IPCC 2007
ii. Universitaet Tübingen (2013, April 23). Iron in primeval seas rusted by bacteria. *ScienceDaily*. Retrieved December 4, 2013, from http://www.sciencedaily.com-/releases/2013/04/130423110750.htm
iii. www.fao.org
iv. J.D. Gilardi, S.S. Duffy, et al. "Biochemical functions of geophagy in parrots: detoxification of dietary toxins and cytoprotective effects," Journal of Chemical Ecology, 25 (4) (1999);897–919.
v. G. Klaus, C. Klaus-Hugi, and B. Schmid, "Geophagy by large mammals at natural licks in the rain forest of Dzanga National Forest Park, Central African Republic," Journal of Tropical Ecology, 14 (1998):829–839.
vi. From www.fao.org/soils-2015
vii. FAO.org
viii. From the World meat markets at a glance (FAO World Food Outlook 2014)
ix. www.epa.gov/rlep
x. Jacobson, Michael F. "More and Cleaner Water." In Six Arguments for a Greener Diet: How a More Plant-based Diet Could save Your Health and the Environment. Washington, DC: Center for Science in the Public Interest, 2006.

Chapter 8

Toxins, Waste, and Pollution

Considering that there are over 7 billion people on the planet, it probably comes as no surprise that we generate exorbitant amounts of waste daily. This waste is causing harm to the existing human population as well as to wildlife and the whole natural world. Waste can take the form of pollution, contaminating the air we breathe, water we drink, and our agricultural soils, causing diseases and threatening the survival of many species and even an entire ecosystem. Toxic chemicals enter the food chain and accumulate in top level predators, decreasing their population size to the point of looming extinction. Plastic waste, both a land and water pollutant, has accumulated in the North Pacific taking form in a mass comparable to the size of Texas! This poses an obvious threat to aquatic ecosystems.

Plastic

Plastic is the by-product of hydrocarbons (organic molecules) derived from petroleum which has been chemically altered so they are durable, flexible, and do not degrade by natural means. They therefore may stick around for hundreds, if not thousands, of years. (Plastic was invented in the 1800s so we still do not have a lot of information on its long-term decomposition.)

Plastics (mostly some types of **PVC**—a popular form of plastic) contain phthalates that are toxic to humans. Phthalates have high human and environmental exposure and have been detected in the human body as well as in food we eat. Phthalates have been linked to developmental problems in the reproductive system (male feminization) and are carcinogenic. Plastics also contain BPA (bisphenol A) and compelling evidence indicates that it mimics estrogen, a sex hormone. This ability to mimic sex hormones can affect reproduction and development. Also of note, BPA has been linked to type II diabetes, obesity, neurological disorders, cancers, increased risk of miscarriages, and may compromise the immune system (all of this especially affects fetuses/infants during developmental stages). Plastics also contain HDPE (high-density polyethylene), the most common resin used in plastic bottles, and it is a suspected carcinogen.

We produce approximately 300 million tons of plastic annually, and only 5% of plastics are effectively recycled while most end up in landfills or the oceans. Keeping this in mind, it should not come as a surprise that in the North Pacific, due to the water currents and movement of gyres, there is an "island" made of plastic comprising almost 10 million square kilometers and is doubling in area each year. It is now estimated that 8 million metric tons of plastic enters the ocean each year.[i] As a result of this, millions of fish, sea turtles, and sea birds die each year from ingesting plastic. A report by the Ellen MacArthur Foundation and presented at the World Economic Forum predicts that by 2050, there will be more plastic than fish by weight. Sun commonly melts plastic and causes plastic residue to dissolve in the oceans. Its effects on the food chain have not been determined. The good news is that all these health and environmental concerns have inspired research and projects that raise public awareness of plastic waste. This has lead to a reduction in the amount of plastic produced, and an increase in the reclamation of plastic and recycling. For instance, you may notice that many plastics are now labeled as recycled or BPA free.

JPL Designs/Shutterstock.com

Sascha Corti/Shutterstock.com

Air Pollution

In 1963, Congress passed the first Clean Air Act to address the escalating pollution problem and to authorize research in order to properly mitigate its adverse effects on the environment. In 1970, Congress passed another Clean Air Act which granted state and Federal governments unprecedented control over the amount of pollution created.

The Clean Air Act of 1970 was seminal because it set standards for ambient air quality and limited pollution emissions of motor vehicles. As a result of the Clean Air Acts, the country has seen a dramatic increase in air quality since 1970.

Rat007/Shutterstock.com

Primary pollution is a pollution that is released directly into the atmosphere either by natural means such as volcanic activity and wildfire dust or by human activities such as automobile and coal-factory pollution. Examples of primary pollutants include sulfur dioxide, carbon monoxide, carbon dioxide, particulate matter, lead, and volatile organic compounds.

These primary pollutants often react with other pollutants in the atmosphere, causing both chemical and physical changes. The substances created by these processes are called **secondary pollutants**. Ozone, nitric acid, and sulfuric acid are examples of secondary pollutants.

Both primary and secondary pollution are harmful to human health and damage the environment. The detrimental effects of both types of air pollution are magnified by the occurrence of **temperature inversions**. A temperature inversion occurs when normal temperature patterns are reversed (warm, lighter air moves over cold, dense air) and cool air is trapped below a column of warm air. Pollution gets trapped below the inversion layer, sometimes for hours, and cannot dissipate easily into the atmosphere as a whole.

Common Outdoor Pollutants

The EPA (Environmental Protection Agency) was created by Richard Nixon's executive order in 1970 in response to first annual Earth Day demonstrations by Gaylord Nelson—protest is important![ii]

The EPA is a Federal organization through which all environmental legislatures move. Enforcement of these regulations is often done on a state level. The EPA has compiled a list of the most common and damaging outdoor pollutants whose concentrations must be controlled. These substances include ozone, VOCs (volatile organic compounds), carbon monoxide, particulates, sulfur dioxide, lead, and nitrogen dioxide. We will touch briefly upon the effects of these pollutants, their impacts on human health and the environment, and why their concentrations must be controlled.

Sai Yeung Chan/Shutterstock.com

Ozone (O_3) is a compound found in high concentrations in the stratosphere (approximately 10–30 miles from earth) and plays a critical role in protecting the earth from UV rays. These UV rays (UVA and UVB—literally cooks your skin, breaks down the structure, so wear your sunscreen!) damage biological molecules such as DNA and compromise our body's cells as well as the cells of the rest of the living world.

Ozone is also found at ground level close to earth's crust. Not naturally found in high concentrations close to earth, ozone is created at ground level when pollution called nitrogen oxides from fossil fuel combustion (via power plants, automobiles, and other forms of organic combustion) react with atmospheric oxygen gas and **VOCs** in the presence of sunlight. Although helpful in the stratosphere, ground level ozone is damaging to both our health and the health of our environment. It weakens the immune system, aggravates heart disease, causes irritation to the nose and lungs, and harms plants. Ozone reacts with sunlight to form a **photochemical smog**, which acts as an irritant to eyes, nose (called a lacrimal response), and to lungs.

VOCs are organic molecules that transform into a gas at room temperature. (Examples are methane, paints, and solvents.) VOCs have been known to cause dizziness, memory lapse, respiratory impairment, and possibly cancer. This is a common phenomenon, especially in the workplace where it may be known as either new building syndrome or sick building syndrome.

Another outdoor pollutant (and also indoor) is **carbon monoxide**, a colorless, odorless gas that is released from car exhaust and forest fires (heaters, furnaces, ranges, etc.). It reacts with red blood cells, disrupting their ability to carry oxygen. It is extremely harmful to human health, takes a long time to exhale from your system, and is possibly deadly.

Particulates are air-borne solids or liquids such as dust, smoke, or even brake-pad debris that accumulate in the air we breathe. Particulates irritate our eyes and respiratory system and can cause bronchitis and asthma. Particulates cause an odd, dark coloration when you blow your nose.

Sulfur dioxide is a colorless gas with a noxious odor released as a result of fossil fuel combustion. It can irritate the cardiovascular and respiratory systems, sometimes causing bronchitis and asthma. (Remember CHNOPS and that fossil fuel used to be a living organism, thus sulfur is a byproduct of combustion.)

Lead is a heavy metal found both naturally and in manufactured goods, but most concentrations in the air are from automobile fuel. Exposure can lead to many health problems, including damage to the nervous system and possible developmental problems (such as female late puberty), and anemia.

Nitrogen dioxide is a highly reactive gas that comes from power plants and motor vehicles. It also contributes to the formation of ground-level ozone. Nitrogen can irritate the eyes, nose, and lungs, and overall respiratory function.

How Do We Reduce Air Pollution?

Cap and trade, sometimes called permit trading, is a practice in which companies receive a limit on the amount of carbon dioxide or other pollution they can release into the atmosphere.

Over time, the limit is gradually reduced, and companies slowly decrease their emissions. The philosophy behind cap and trade is that the companies will slowly learn to cut down on their amount of pollution and improve overall environmental quality. If companies are under their limit or cap, and release less pollution than allowed, they receive **pollution credits**. These pollution credits may be used as a currency that companies can sell to other companies for money. The purchase of these credits allows the buyer to produce more pollution than their limit would otherwise allow.

Cap and trade has been considered successful in controlling sulfur dioxide, a pollutant which causes acid rain. But critics are dubious about whether it would be effective in controlling major pollutants such as carbon dioxide. Critics express concern that this system of cap and trade leaves too much room for dishonesty and corruption. Companies may fudge or embellish data about their emissions in order to receive unearned credits. Another common criticism of this practice is its underlying philosophy. To quote Albert Einstein, "We can't solve problems by using the same kind of thinking we used when we created them." In other words, the practice of cap and trade is a system that uses money as the ultimate bottom line—and is therefore the same kind of thinking that generated the pollution problem at the start.

Another possible solution to the air pollution problem is to tax companies based on how much pollution they release. This is the Polluter Pays idea, making companies accountable for their own waste so they evaluate existing practices to avoid paying high tax rates.

Another solution is to offer government subsidies and tax breaks to companies that are environmentally friendly, providing incentives to companies that are more environmentally responsible. Canada has recently imposed a carbon tax which has been pretty successful in curbing emissions. One example of this approach is shown by many companies that are buying and planting trees in order to offset their carbon footprint.

Indoor Pollution

Radon gas is a radioactive element that is found naturally in rock and soil. It is known to leak into homes through cracks or other openings in the foundation of homes and infrastructures. Radon gas is second to smoking for causes of lung cancer (a possible solution is to place plastic covers on house floors in areas that have high concentrations of uranium deposits). Other common sources of indoor pollution may be poorly ventilated stoves, tobacco smoke, and even the burning of wood and coal, all of which release particulate matter, nitrogen oxides, and carbon monoxide, to name a few. Tobacco smoke contains cancer-causing carcinogens. When homes are not ventilated properly, which is often the case especially in developing countries, the pollution is stagnant and is breathed unknowingly by the homes' occupants. For the above reasons, the Environmental Protection Agency considers indoor pollution the most dangerous type of pollution that exists today.

Soil Pollution

Millions of tons of trash are generated each year and placed into landfills. To prevent environmental contamination, the federally regulated landfills designate areas for disposal of hazardous waste. Such hazardous waste is produced often, believe it or not, from everyday household items.

By EPA standards, hazardous waste is classified as anything that is capable of causing harm, death, or illness to humans when released into the environment. Hazardous substances are typically corrosive, combustible, or toxic. **Corrosive items** are reactive chemicals that damage or compromise the surface of a material. Strong acids such as sulfuric acid, or battery acid, are examples of corrosive substances. **Combustible items** have a strong capability of igniting or burning at normal temperatures. Ethanol, solvents, and paint thinners are often combustible. **Toxic chemicals** may cause either long- or short-term harm to humans and animals when they are exposed to such chemicals. Examples of toxic chemicals are lead, mercury, and asbestos.

Landfills that are improperly managed or deteriorating may cause soil contamination when waste leaks out from the confines of the landfills. Soil can also be contaminated from chemicals and pollutants that get caught up in runoff. These chemicals often come from factories and agriculture farms. Some contaminants, such as BPA (commonly found in wastewater) may act as endocrine disruptors that compromise hormone production. Microorganisms found in wastewater are also known soil contaminants. Up to 25 million gallons of oil is spilled annually, killing plants and wildlife, among innumerable adverse impacts on human health and the environment.

DDT (dichlorodiphenyltrichloroethane) was a pesticide used in the 1960s and 1970s until it was banned in 1972. DDT was enormously effective in controlling the mosquito population. In the ecosystem, DDT may take up to 15 years to break down, so it has high **persistence** (amount of time it takes for a substance to break down). However, insects became resistant to DDT, and its efficacy decreased. Its toxicity to fish became increasingly problematic as well.

As it entered the food chain, it affected not just fish, but many birds such as the brown pelican and the American bald eagle. It compromised how birds use their body's calcium stores, resulting in thinner egg shells that could not sustain an incubating bird—the weight of the parent would crush the egg. Bald eagles became endangered, but aggressive preservation efforts proved successful and they were removed from the federal endangered species list in 2007. But the damage did not end there. DDT is a toxic chemical whose side effects in humans include cancer, liver problems, reproductive

FloridaStock/Shutterstock.com

problems, and damage to the nervous system. It is still used in many tropical countries to control the spread of mosquitos.

The Montrose Chemical Corporation, a Los Angeles based company, started using DDT after World War II. Despite warnings and known health risks, Montrose did not stop using DDT until the early 1980s. It is estimated that they discharged over 1,500 tons of DDT waste into the Pacific where it settled in ocean sediment.

Many pesticides and other chemicals, such as DDT, do not degrade over time (high persistence) and are often fat soluble. These pesticides **bioaccumulate** in the food chain and are found in high concentrations in predators high up in the food chain. For this reason, toxic chemicals such as mercury and DDT are found in high concentrations in large predatory fish. This phenomenon is called **biomagnification**.

Water Pollution

Water that blankets the earth takes on various forms—from fresh mountain springs to underground aquifers, from saltwater to watersheds to wetlands. All these sources are major components of the hydrologic cycle. The Clean Water Act of 1948 was designed to control the pollution and contamination of U.S. water. Although strict measures are taken to procure clean water for our growing population, 40% of freshwater in the United States is still too contaminated for fishing or swimming. In 1974, the Safe Drinking Water Act was passed to ensure and protect the quality of drinking water—mostly from underground sources.

Some common biological water contaminants include pathogenic microorganisms, such as protozoa (dysentery), bacteria (such as cholera and fecal coliform bacteria), and viruses (polio and hepatitis). Natural materials such as sediments also act as contaminants. Sediments are nonbiological material derived from the weathering of rocks. Water pollutants also include mining wastes, such as acids, mercury, cyanide, and arsenic. Chemical disinfectants such as chlorine, sewage, runoff from factories and croplands, by-products from fire retardants, electronics, medicines, and corrosive household plumbing—all leach into our lakes and streams, making them unsafe for human use. They can also alter the pH of water (a measure of the buffer capacity of water), killing many organisms that have narrow habitable ranges. Methane from cow waste is another major cause for water pollution, and often ends up in bodies of water causing methane lagoons.

Pollution emitted directly into a surface through a confined, direct means such as a pipe, drain, or sewage is called **source pollution**, since it can be traced directly to its source. **Nonsource** pollution is water pollution that is not categorized as source pollution. Nonsource pollution tends to be more diffuse, less concentrated, and widespread. As runoff moves through a watershed, it picks up pollutants and deposits them into lakes and rivers, as well as underground sources of drinking water and surface water. There is no single point at which it is released, so it is considered nonsource pollution. Nonsource pollution can be sand, clay, and silt, but also human-made toxins.

Other forms of water pollution can be in the form of noise pollution—which can affect the sonar, communication, and life cycles of marine animals. In 1972, the Noise Control Act was passed. Enforced on a national level, it serves to protect the health and well-being of people, wildlife, and the environment by promoting a noise-free environment. Thermal pollution (heat) from global warming and power plants can also affect aquatic life, and so is also considered a form of water pollution.

Source Pollution

Chemicals and contaminants in groundwater (from http://water.usgs.gov/edu/groundwater-contaminants.html)

Inorganic Contaminants Found in Groundwater		
Contaminant	**Sources to groundwater**	**Potential health and other effects**
Aluminum	Occurs naturally in some rocks and drainage from mines.	Can precipitate out of water after treatment, causing increased turbidity or discolored water.
Antimony	Enters environment from natural weathering, industrial production, municipal waste disposal, and manufacturing of flame retardants, ceramics, glass, batteries, fireworks, and explosives.	Decreases longevity, alters blood levels of glucose and cholesterol in laboratory animals exposed at high levels over their lifetime.
Arsenic	Enters environment from natural processes, industrial activities, pesticides, and industrial waste, smelting of copper, lead, and zinc ore.	Causes acute and chronic toxicity, liver and kidney damage; decreases blood hemoglobin. A carcinogen.
Barium	Occurs naturally in some limestones, sandstones, and soils in the eastern United States.	Can cause a variety of cardiac, gastrointestinal, and neuromuscular effects. Associated with hypertension and cardiotoxicity in animals.
Beryllium	Occurs naturally in soils, groundwater, and surface water. Often used in electrical industry equipment and components, nuclear power and space industry. Enters the environment from mining operations, processing plants, and improper waste disposal. Found in low concentrations in rocks, coal, and petroleum and enters the ground and	Causes acute and chronic toxicity; can cause damage to lungs and bones. Possible carcinogen.

Cadmium	Found in low concentrations in rocks, coal, and petroleum and enters the groundwater and surface water when dissolved by acidic waters. May enter the environment from industrial discharge, mining waste, metal plating, water pipes, batteries, paints and pigments, plastic stabilizers, and landfill leachate.	Replaces zinc biochemically in the body and causes high blood pressure, liver and kidney damage, and anemia. Destroys testicular tissue and red blood cells. Toxic to aquatic biota.
Chloride	May be associated with the presence of sodium in drinking water when present in high concentrations. Often from saltwater intrusion, mineral dissolution, industrial and domestic waste.	Deteriorates plumbing, water heaters, and municipal water-works equipment at high levels. Above secondary maximum contaminant level, taste becomes noticeable.
Chromium	Enters environment from old mining operations runoff and leaching into groundwater, fossil-fuel combustion, cement-plant emissions, mineral leaching, and waste incineration. Used in metal plating and as a cooling-tower water additive.	Chromium III is a nutritionally essential element. Chromium VI is much more toxic than Chromium III and causes liver and kidney damage, internal hemorrhaging, respiratory damage, dermatitis, and ulcers on the skin at high concentrations.
Copper	Enters environment from metal plating, industrial and domestic waste, mining, and mineral leaching.	Can cause stomach and intestinal distress, liver and kidney damage, anemia in high doses. Imparts an adverse taste and significant staining to clothes and fixtures. Essential trace element but toxic to plants and algae at moderate levels.
Cyanide	Often used in electroplating, steel processing, plastics, synthetic fabrics, and fertilizer production; also from improper waste disposal.	Poisoning is the result of damage to spleen, brain, and liver.
Dissolved solids	Occur naturally but also enters environment from man-made sources such as landfill leachate, feedlots, or sewage. A measure of the dissolved "salts" or minerals in the water. May also include some dissolved organic compounds.	May have an influence on the acceptability of water in general. May be indicative of the presence of excess concentrations of specific substances not included in the Safe Water Drinking Act, which would make water objectionable. High concentrations of dissolved solids shorten the life of hot water heaters.
Fluoride	Occurs naturally or as an additive to municipal water supplies; widely used in industry.	Decreases incidence of tooth decay but high levels can stain or mottle teeth. Causes crippling bone disorder (calcification of the bones and joints) at very high levels.

(Continued)

Inorganic Contaminants Found in Groundwater (*Continued*)		
Contaminant	**Sources to groundwater**	**Potential health and other effects**
Hardness	Result of metallic ions dissolved in the water; reported as concentration of calcium carbonate. Calcium carbonate is derived from dissolved limestone or discharges from operating or abandoned mines.	Decreases the lather formation of soap and increases scale formation in hot-water heaters and low-pressure boilers at high levels.
Iron	Occurs naturally as a mineral from sediment and rocks or from mining, industrial waste, and corroding metal.	Imparts a bitter astringent taste to water and a brownish color to laundered clothing and plumbing fixtures.
Lead	Enters environment from industry, mining, plumbing, gasoline, coal, and as a water additive.	Affects red blood cell chemistry; delays normal physical and mental development in babies and young children. Causes slight deficits in attention span, hearing, and learning in children. Can cause slight increase in blood pressure in some adults. Probable carcinogen.
Manganese	Occurs naturally as a mineral from sediment and rocks or from mining and industrial waste.	Causes aesthetic and economic damage, and imparts brownish stains to laundry. Affects taste of water, and causes dark brown or black stains on plumbing fixtures. Relatively nontoxic to animals but toxic to plants at high levels.
Mercury	Occurs as an inorganic salt and as organic mercury compounds. Enters the environment from industrial waste, mining, pesticides, coal, electrical equipment (batteries, lamps, switches), smelting, and fossil-fuel combustion.	Causes acute and chronic toxicity. Targets the kidneys and can cause nervous system disorders.
Nickel	Occurs naturally in soils, groundwater, and surface water. Often used in electroplating, stainless steel and alloy products, mining, and refining.	Damages the heart and liver of laboratory animals exposed to large amounts over their lifetime.
Nitrate (as nitrogen)	Occurs naturally in mineral deposits, soils, seawater, freshwater systems, the atmosphere, and biota. More stable form of combined nitrogen in oxygenated water. Found in the highest levels in groundwater under extensively developed areas. Enters the environment from fertilizer, feedlots, and sewage.	Toxicity results from the body's natural breakdown of nitrate to nitrite. Causes "bluebaby disease," or methemoglobinemia, which threatens oxygen-carrying capacity of the blood.
Nitrite (combined nitrate/nitrite)	Enters environment from fertilizer, sewage, and human or farm-animal waste.	Toxicity results from the body's natural breakdown of nitrate to nitrite. Causes "bluebaby disease," or methemoglobinemia, which threatens oxygen-carrying capacity of the blood.
Selenium	Enters environment from naturally occurring geologic sources, sulfur, and coal.	Causes acute and chronic toxic effects in animals—blind staggers" in cattle. Nutritionally essential element at low doses but toxic at high doses.

Silver	Enters environment from ore mining and processing, product fabrication, and disposal. Often used in photography, electric and electronic equipment, sterling and electroplating, alloy, and solder. Because of great economic value of silver, recovery practices are typically used to minimize loss.	Can cause argyria, a blue-gray coloration of the skin, mucous membranes, eyes, and organs in humans and animals with chronic exposure.
Sodium	Derived geologically from leaching of surface and underground deposits of salt and decomposition of various minerals. Human activities contribute through de-icing and washing products.	Can be a health risk factor for those individuals on a low-sodium diet.
Sulfate	Elevated concentrations may result from saltwater intrusion, mineral dissolution, and domestic or industrial waste.	Forms hard scales on boilers and heat exchangers, can change the taste of water, and has a laxative effect in high doses.
Thallium	Enters environment from soils; used in electronics, pharmaceuticals manufacturing, glass, and alloys.	Damages kidneys, liver, brain, and intestines in laboratory animals when given in high doses over their lifetime.
Zinc	Found naturally in water, most frequently in areas where it is mined. Enters environment from industrial waste, metal plating, and plumbing, and is a major component of sludge.	Aids in the healing of wounds. Causes no ill health effects except in very high doses. Imparts an undesirable taste to water. Toxic to plants at high levels.

Organic Contaminants Found in Groundwater

Contaminant	Sources to groundwater	Potential health and other effects
Volatile organic compounds	Enter environment when used to make plastics, dyes, rubbers, polishes, solvents, crude oil, insecticides, inks, varnishes, paints, disinfectants, gasoline products, pharmaceuticals, preservatives, spot removers, paint removers, degreasers, and many more.	Can cause cancer and liver damage, anemia, gastrointestinal disorder, skin irritation, blurred vision, exhaustion, weight loss, damage to the nervous system, and respiratory tract irritation.
Pesticides	Enter environment as herbicides, insecticides, fungicides, rodenticides, and algicides.	Cause poisoning, headaches, dizziness, gastrointestinal disturbance, numbness, weakness, and cancer. Destroys nervous system, thyroid, reproductive system, liver, and kidneys.
Plasticizers, chlorinated solvents, benzo[a] pyrene, and dioxin	Used as sealants, linings, solvents, pesticides, plasticizers, components of gasoline, disinfectant, and wood preservative. Enters the environment from improper waste disposal, leaching runoff, leaking storage tank, and industrial runoff.	Cause cancer. Damages nervous and reproductive systems, kidney, stomach, and liver.

Microbiological Contaminants Found in Groundwater		
Contaminant	**Sources to groundwater**	**Potential health and other effects**
Coliform bacteria	Occur naturally in the environment from soils and plants and in the intestines of humans and other warm-blooded animals. Used as an indicator for the presence of pathogenic bacteria, viruses, and parasites from domestic sewage, animal waste, or plant or soil material.	Bacteria, viruses, and parasites can cause polio, cholera, typhoid fever, dysentery, and infectious hepatitis.

Some information on this page is from Waller, Roger M., Ground Water and the Rural Homeowner, Pamphlet, U.S. Geological Survey, 1982

How Do We Deal with Water and Soil Pollution?

Nanotechnology is one method used to prevent and to clean up water and soil pollution. Nanotechnology involves manipulating the environment on a molecular level. Current research suggests that nanotechnology is efficient in removing contaminants like chlorine from the soil; a fungus from the genus *Pestalotiopsis* has recently been discovered in the Amazon that is able to break polyurethane (a material found in rubber and plastic, and has very high persistence) down in anaerobic (no oxygen) conditions. Anaerobic conditions are similar to those at the bottom landfills. This is a breakthrough since many bacteria used in bioremediation require oxygen, and the study suggests that the fungus may be useful to break plastic down in landfills.[iii]

Green infrastructure is an overall approach to community living that integrates nature into developed communities. The infrastructure mimics natural processes and uses vegetation and soil to capture and enhance the hydrologic cycle to prevent excess runoff in urban areas.

Green technology is a means that uses natural organisms to clean up pollution.

Bioremediation is a form of both nanotechnology and green technology that uses microorganisms such as bacteria and protozoa to break down toxic chemicals in the soil and biotransform them into CO_2.

Phytoremediation is a green technological practice used to remove toxic chemicals from the soil through plants, since plants have the ability to absorb heavy metals through their roots. A benefit of phytoremediation is localizing toxic chemicals in a controlled plant population. But care must be taken that

Dr. Morley Read/Shutterstock.com

Bioremediation Pond for Soil Contaminated with Crude Oil

they stay localized so plants do not hybridize with natural plant populations. Plants must later be removed in order to remove the heavy metals and other contaminants.

Waste Management

Each year, Americans generate millions of tons of trash from our everyday activities, whether we are aware of this or not. Almost half of all waste is considered either *agricultural* or *industrial* waste that comes from manufactures goods or the food we eat. Mining for ore and metals creates *mining waste.* And apartments, homes, small business, colleges, dormitories create *municipal solid waste.*

So how do we safely dispose of all our household and industrial items while maintaining ecosystem health—preventing air, soil, and water pollution? Long gone are the days when we used to drop our trash off at the open dump. These days, waste disposal is a lot more controlled and systematic. Thanks to stricter laws, our waste is now very carefully regulated so minimum damage is inflicted on the environment.

Most of our solid household waste goes into federally regulated landfills that contain areas designated for disposal of municipal solid waste. Such waste is monitored to ensure it is safe for disposal in our municipal landfills. Some of the items that get disposed are construction debris, industrial solid waste, municipal solid waste, and nonhazardous sludge. Hazardous household items, not safe for disposal, have their own drop-off site. Industrial wastewater that is contaminated with toxic chemicals cannot be disposed of by regular means, because it can contaminate groundwater and leak into the water table—so it is injected into deep wells. This sometimes leads to induced or man-made earthquakes.[iv]

Huguette Roe/Shutterstock.com

Bulldozer Working on a Landfill

Landfills are specifically selected for disposal in nonenvironmentally sensitive areas, which means that building on wetlands, floodplain zones, or fault lines, is prohibited (to prevent the aforementioned earthquakes).

Landfills are specifically designed and engineered to prevent pollution from leaching into the soil and contaminating the water table. This means that the site must have certain geological requirements such as a bed of clay soil lining the site (to protect groundwater from being contaminated). Landfills are constantly monitored for quality assurance, and groundwater is tested to make sure it is not contaminated. Landfills are periodically covered with soil to prevent odor and to prevent infestations from insects and rodents.

The EPA has outreach programs affiliated with some landfills to capture methane gas and carbon dioxide and eventually convert them into energy. This keeps carbon emissions down, improves air quality, and is a way of harnessing natural gas so that it is renewable (see Chapter 5).

According to the Environmental Protection Agency, of the material and trash that we throw out here in the United States, roughly 50% is discarded all together (meaning it ends up in landfills), 13% is combusted for energy (meaning it is incinerated), and 34% is composted or recycled. From this, we conclude that half of our waste is still being discarded! Landfills are filling up quickly as a result of our overwhelming generation of waste and refuse. Suitable sites for landfills are becoming increasingly scarce.

Logically, many people argue that we need to lessen our dependency on landfills and incinerators altogether. This happens when we cut down on the amount of trash and waste that we create and ramp up the amount of recycling. If we profoundly cut down on the amount of trash, we would also reduce our use of natural resources such as water, timber and minerals. We would also significantly reduce the amount of greenhouse gases and pollution associated with extraction of these natural resources.

We achieve less waste with the three 'Rs': Let us *Reduce, Re-use* and *Recycle* our waste.

Reduce: Reduce the amount of waste we create. Buy in bulk. Borrow or rent material that is not used often. This may include printers, ladders, bicycles, etc. This is also a great way to save money!

Re-use: Buy repurposed material or used goods like clothing, silverware, and building material. Buy products with less packaging. Repair or get maintenance done on appliances and electronics rather than buying them new. Donate your unwanted items and electronics to religious institutions, charities or thrift stores.

Recycle: Collection for recycling happens curbside or at deposit centers that offer refunds. Contamination can be high in recycle bins, so please take care that material is clean and sorted.

When buying goods and products at the market, look for phrases such as: 'recycled-content product,' 'post-consumer product,' or 'recyclable product.' Many common household items such as paper towels, newspapers, nails, trash bags, cans, plastic bottles, aluminum and glass are increasingly being made with recycled material. The best way to support recycling is to vote with your dollar, purchase items that can be recycled or are made from recycled material.

Review Questions

1. Distinguish between primary and secondary pollution. Please provide an example of each.

2. What is a temperature inversion, and how does it contribute to air pollution?

3. Name 6 common outdoor pollutants whose concentrations are controlled by the EPA. Why is each of these harmful and in need of regulation?

4. What does it mean when a chemical is considered to have high "persistence." Please provide an example.

5. Please provide examples of "bioremediation" and "phytoremediation."

Discussion Questions for Class

A. How have the different types of pollution contributed to earth's environmental problems?

B. How have ambient air and air quality standards improved since the EPA passed the air quality acts?

C. Can you suggest ideas that would offer companies an incentive to reduce pollution?

D. What types of disposable items are found in the Pacific trash gyre? Which are most deadly to the endemic animal populations?

SII/Shutterstock.com

E. Can you think of ways to contribute to green infrastructure?

F. Should carbon dioxide be regulated by the EPA along with all the other outdoor pollutants (ozone, particulates, etc.)?

G. Why is radon gas considered natural when it is so harmful?

H. What causes new building syndrome and how is this corrected?

I. Why does the EPA consider indoor air pollution the most dangerous?

J. How are landfills different than open dumps? Why are they better for the environment?

Your Personal Sustainability Journal

Reduce, Re-use, and Recycle.

Furnish your home with plants such as the purple waffle plant, English ivy, and the Asparagus fern. They remove VOCs and other indoor pollutants.

Recycle plastics and use biobags that feel like plastic but are made from plant and fossil raw material and are 100% compostable. There are also bags available made of corn husks.

Store hazardous material carefully to avoid spillage and leakage and follow directions for proper disposal.

Check the recycling symbol on your plastic bottles to make sure they do not contain chemicals that will leach into the soil.

When possible, store water or food in glass containers rather than plastic containers. If you are using plastic, then keep it out of direct sunlight.

Consider that 20 billion diapers are thrown out each year. They take hundreds of years to biodegrade. If you plan to start a family, look into using cloth diapers, or re-usable 'hybrid' diapers with flushable, biodegradable inserts.

Add your own suggestions here.

Endnotes

i. https://www.ellenmacarthurfoundation.org/publications/the-new-plastics-economy-rethinking-the-future-of-plastics-catalysing-action

ii. http://www.epa.gov/earthday/history.htm

iii. Biodegradation of Polyester Polyurethane by Endophytic Fungi, from *Applied and Environmental Microbiology*; Russell, Huang, et al.

iv. http://www.usgs.gov/newsroom/article.asp?ID=4202#.VUf80flVikp

Chapter 9

Water! Aquatic Ecosystems, Fisheries, Aquaculture, and Water Purification

DID YOU KNOW?

In desert ecosystems, all plants and wildlife have their own unique adaptations for conserving water. In order to survive hot, dry desert climates, camels have adapted the ability to conserve and store copious amounts of water for prolonged periods of times. When given the opportunity, camels drink up to one-third of their body weight of water. They store their water reserves as fat—in what we know as the camel hump. Fat in their humps store metabolic water—waste from cellular respiration. When water is needed, the fat stored in the hump is broken down into water. But camel adaptations do not stop there—during peak sunlight hours when temperatures are most extreme, camels face directly to the sun to limit bodily exposure to the hot sun. They have oval-shaped red blood cells to maintain consistent blood flow during dry spells and prolonged periods of dehydration. Ironically, their thick fur raises their body temperature which in turn reduces water loss.

hainaultphoto/Shutterstock.com

The Namib beetle of Africa (found in the Namib Desert of the Atlantic coast) uses its back as a drinking cup of sorts. The beetle angles its body so fog or water droplets accumulate on its back and then strategically angles its body so water droplets enter its mouth, and the organism successfully hydrates itself.

ChameleonsEye/Shutterstock.com

A Namib Desert Beetle

Water in the Ecosystem

In Chapter 2, we discussed the unique properties of water. To briefly review, water is a polar molecule whose unique properties include cohesion, adhesion, surface tension, high specific heat, and high heat of vaporization. Water is the only substance known today considered a universal solvent or close to it, since it will react with anything containing a charge. Water is also the only substance found commonly in a solid, liquid, and gaseous form.

Since all organisms need water, living beings in a given area are mutually linked by a **watershed**, an aquatic network where local surface water drains and eventually effuses into a single point or waterway.

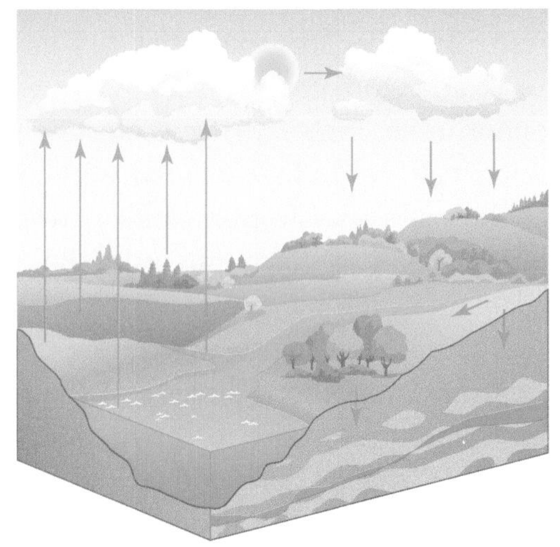
MarinaMariya/Shutterstock.com

The Hydrologic Cycle

When rain and precipitation come down to earth, water percolates into the ground until it hits a saturation point—usually when it has seeped to rocky soil. At this point, the water absorbed by the ground has become **groundwater**.

The **water table** is the point at which the pressure of groundwater equals atmospheric pressure. The water table is often observed as a flat or level surface of water. When precipitation does not get fully absorbed by the ground, it is lost as **runoff** as it moves downhill from upper to lower altitudes. With it, runoff can take the first few inches of nutrient-rich ground soil called **topsoil**. The flow of water strips away topsoil causing soil erosion, which can lead to

desertification (when soil can no longer hold nutrients). Desertification often means that vegetation can no longer grow and the soil is rendered useless. Desertification is responsible for the Dust Bowl in 1930s in Oklahoma, the agricultural catastrophe that caused over 50% of the local population to relocate. (It is also important to note that dry soil, monoculture agricultural practice, and high winds were mostly responsible for the dust bowl. The end result was desertification.)

Pi-Lens/Shutterstock.com

Hail, Fall Leaves, and Debris Block up a Sewer Hole, Restricting Runoff Flow

Aquatic Habitats

There are many different aquatic habitats, but all are categorized under the umbrella of either **freshwater** or **marine** habitats. Marine habitats contain high salinity (oceans for example) whereas freshwater do not (freshwater lakes and streams for example).

There are three major **oceans**, the Atlantic, the Pacific (the largest and the deepest), and the Indian. The average depth of oceans is approximately 2 miles! The depths of the oceans contain the longest mountain ranges and active volcanos. Oceans are a marine habitat and can be broken down into zones, depending on depth and amount of sunlight received. Each zone has its own unique aquatic life with adaptations to ocean depth, pressure, and amount of available sunlight. Furthermore, the ocean is its own complex ecosystem driven by the relationship between carbon, oxygen, and other nutrients.

Dominic Gereri/Shutterstock.com

The **euphotic zone** sits atop a continental shelf and receives the most net sunlight of all ocean zones (80% of solar energy is found in the first 10 meters of the ocean). Since this zone receives plenty of sunlight, the aquatic life will take advantage and use it as an energy source. Many photosynthetic organisms such as algae and phytoplankton are found here, but nonphotosynthetic organisms such as zooplankton (which feed on algae) are also found in this zone, as well as small fish. Photosynthetic phytoplankton, similar to trees and plants found on land, are instrumental to the ecosystem functioning. It absorbs and assimilates carbon and nutrients. These are processed and expelled as organic matter that eventually reaches the oceans' subsurface. In the tropical climates, the euphotic zone contains the coral reefs, home to sea creatures such as anemones and sponges. The **photic** zone is the upper layer of the open seas (not on the continental shelf). Whales are often found at this layer as they break the water surface to obtain oxygen.

The **bathyal zone**, 700–1,000 meters down, contains cooler temperatures (roughly 5°C to 10°C) and very little sunlight. The type of marine life found here is classified as **nekton**—strong swimmers such as sharks and whales. Squids and bioluminescent fish are also found in this layer. They are nonphotosynthetic organisms and therefore heterotrophs.

In the **abyssal zone**, 2,000–4,000 meters down, we find cooler temperatures (less than 5°C) high pressure, and no light. One finds animals adapted to these conditions such as sea stars, squid, and eels in this zone. One also finds animals that rely on bioluminescence for communication, predation, and to attract mates.

Even further down, in oceanic trenches, is a region called the **hadal zone**. Hadal zones are rich with organisms that feed on chemicals such as hydrogen sulfides released from ocean vents. The process in which organisms obtain nourishment from such chemicals is called **chemosynthesis**. The deepest parts of the oceans contain more biodiversity and biodensity than even the tropical rainforests. The hadal zone contains ocean sediment and bacteria that can live in temperatures as extreme as 180°C. It is also home to jellyfish, bottom dwellers, and such curious creatures as tubeworms and sea cucumbers. (Tubeworms are especially curious because they have no mouth or digestive system. They feed on the bacteria and convert it to sugar.)

plampy/Shutterstock.com

Our oceans take up roughly 70% of the earth's surface, yet they are all being degraded in some form or another through human activity. Agricultural runoff, point source pollution, nonpoint source pollution, overfishing, invasive species, acidification, plastic, ghost nets, changing water levels, climate change, and mass transportation are all harming the oceans. In fact, disturbance is so extensive, that it is estimated that 20%–25% of common or well-known fish species are at risk of extinction.[i] Many marine biologists and scientists feel that threats to marine ecosystems are not taken as seriously as threats to terrestrial ecosystems, even though both have comparable risks and extinction rates. Perhaps this may be explained through realizing that we know so little about the oceans. This would be remedied if people would shift their conservation efforts to include prioritizing marine ecosystems. Integration between marine and coastal management would hopefully curb the amount of disturbance and ensure sustainable use of the ocean's biodiversity.

Corals evolved approximately 600 million years ago, around the time multicellular animals first started to evolve. **Coral reefs** are found in shallow temperate marine habitats that receive plenty of sunlight. Here, sea anemone-type creatures called polyps secrete calcium bicarbonate that creates the corals on the shallow ocean floor. This becomes home to many species of algae, small fish, and other marine life. Their skeletons provide home for other polyps and algae, enabling algae to access to the sun for photosynthesis. Thus, the corals forge a mutualistic symbiotic relationship with the algae (the photosynthetic algae provide nourishment for the coral animals) to create the beautiful colorful hues of the coral reefs.

Vlad61/Shutterstock.com

The Brilliant Colors of the Coral Reefs

The coral reefs have the most biodiversity of any marine habitat. It is comprised of sea urchins, shrimps, sponges, sea turtles, whales, and dolphins, to name a few. Approximately 4,000 fish inhabit undisturbed coral reefs, and it is often coined the tropical rainforest of the oceans.

Ecologically, the reefs are significant because not only do they contain the most biodiversity of any aquatic ecosystem, but they are also biologically productive and act as buffer zones for storms.

Polyps (the coral animal) act as a **foundation species** by creating the reefs, and research suggests they can affect local climates as well. When temperatures rise, these polyps release a sulfur compound that increases the amount of local cloud cover. Such cloud cover protects this fragile ecosystem from UV light and thus moderates the temperature. (It has long been known that certain types of algae had this capability, but only recently discovered that the coral animal has it as well.[ii])

The coral reefs also contribute other value to humans, such as providing us with pharmaceuticals, seafood, ecotourism, as well as overall beauty. Thus, there is great incentive to conserve the reefs.

Despite providing immeasurable value, the coral reefs face great disturbance, not only from natural phenomena such as storm activity, but also from human-influenced activity as well. Such human activity includes the introduction of predatory invasive species such as starfish that attack and harm the corals.

Australia's Great Barrier Reef, for example, one of the seven wonders of the natural world as well as one of the only living structures visible from space, is also the largest animal-made creation on earth. With such auspicious notoriety, it is no wonder that the Great Barrier Reef is a major tourist destination—but not without huge sacrifice to its health. It is estimated that 60% of the Great Barrier Reef is being degraded, mostly through human activity. It is also estimated that the predatory starfish is responsible for the loss of half of the corals in Australia's Great Barrier Reef.

Other human-influenced disturbances include the effects of global warming and overharvesting of kelp and various fish species. To catch live coral fish, divers place cyanide tablets in plastic bottles and squirt the mixture into crevices in order to stun and capture coral reef fish (that are typically used by humans for tropical aquariums). Not only is cyanide dangerous to humans and fish but also lethal to the corals. And even more disturbances to the coral reefs come from fertilizers and agricultural runoff.

Coral bleaching is a phenomenon that occurs when corals, under duress, expunge the brilliantly colored algae and thus lose their brilliant coloring. Coral bleaching is the result of change in water temperature and pH among other variables and is a signal that coral reefs are under stress. After corals expunge the algae during bleaching, they start feeding carnivorously and then eventually die prematurely, or they get overtaken by algae, suffocate and die.

Intertidal zones, or sandy beaches, are also a shallow-water marine habitat, and are very dynamic. These changes are a result of the magnetic pull from the sun and the moon. Because of the changing ecosystem variables such as light, temperature, salinity, and the force of pounding waves, they are an ideal place to study ecology. Some of the marine life found in this ecosystem are amphibious and live on both land and in the ocean. Sea life found here are adapted to burrowing into the sand or under rocks to withstand the changing tides, water levels, and temperatures. They also have adaptations that help them anchor and protect from water loss. This zone has more salinity than other marine ecosystems because of high evaporation rates and so the marine life found here are adapted to withstand such dramatic changes. Some examples of marine life found here include sea stars, mussels, seaweed, hermit crabs, and barnacles.

Thermal pollution (as a result of climate change), sediment, runoff, and other types of pollution have been affecting intertidal zones. The heavy amount of human foot traffic negatively impacts the ecosystem; human encroachment due to visitation and exploration of tidepools causes disturbance, as does kelp harvesting. Other human activity generates trash and pollution such as cigarette butts, food wrappers, and other various types of waste.

Patches of land drain their groundwater as runoff which becomes concentrated as **rivers and streams**, which serve as freshwater aquatic habitats. The outer banks of rivers and streams (**riparian zones**) contain rich, moist soil, and produce lush vegetation. Rivers and streams have diverse ecological functions that include, but are not limited to: renewing oxygen supply to fish and freshwater inhabitants; removing waste; and delivering food to all living organisms connected through the river system and network. For humans, they are significant because they provide transportation and waste disposal, and they irrigate land. Riparian zones and fresh running water are often polluted as a result of over-grazing of livestock. Methane and other waste contaminates streams and riparian areas, compromising the sensitive species that dwell in these areas.

Mariusz S. Jugielewicz/Shutterstock.com

Lakes are basins or depressions in land that collect and act as a storehouse of water from runoff, mountain glaciers, lakes, or streams. They can be natural or manmade. Lakes are usually freshwater but may have high concentrations of salt, as does the Great Salt Lake. Desert lakes typically have high salinity while alpine (mountainous) lakes are more dilute. **Oligotrophic** lakes have an ample supply of oxygen while **eutrophic** lakes have low oxygen supply. Eutrophic lakes are usually rich in nutrients, organic material, and minerals. Because they have an ample supply of nutrients, eutrophic lakes typically facilitate a lot of natural processes and functions such as

daniel towia/Shutterstock.com

Intertidal Zone/Rocky Shore

Thomas Barrat/Shutterstock.com

An Alpine Lake

photosynthesis and decomposition—that uses up most of the oxygen supply. Oligotrophic lakes with their ample oxygen concentrations usually have comparatively less ecological processes.

Just as the vast, massive oceans are characterized by different layers, so are the still-water ecosystems, (or lakes) that are relatively large in size. Lakes have three distinct zones, and temperature can often change dramatically with depth. This temperature variation is called **thermal stratification.**

The **littoral zone** is near the edge of a lake or pond, where aquatic life resides in shallow water. Reeds, algae and cattails are all examples of life in the littoral zone.

The **limnetic zone** is the upper layer of open water—but not found on the edge of the lake or pond. High amounts of sunlight seep into the limnetic zone, and there is an abundance of photosynthetic organisms that are adapted to getting nutrients from the sun (such as phytoplankton). Larger fish also dwell in this zone.

The **profundal zone** is beneath the limnetic zone at the bottom of the lake, and does not receive an abundance of sunlight. This means that there are no plants, algae, or other photosynthetic organisms. Detritus falls from the upper layers into the profundal zone and bacteria break down and recycle into the ecosystem. Just below the profundal zone is the **benthic zone**, or the lake bottom.

Point source pollution is a threat to lakes, polluting it with human waste and toxic chemicals. Nonpoint source pollution is also a threat. Nonpoint source pollution includes runoff, agricultural waste, fertilizers and pesticides, and salt and sediment. Atmospheric pollution such as smoke, acid rain, and thermal pollution from climate change are all negatively impacting our lakes as well.

Because they are contained bodies of water with no outlet, lakes are vulnerable to invasive species. For instance, the lamprey, a type of sucker fish, was accidently introduced into Lake Erie and subsequently killed off a high percentage of its rainbow trout and whitefish population. (Refer to Chapter 3.)

Estuaries are found where freshwater meets saltwater, or more specifically, where rivers empty into the ocean. Estuaries have fluctuating salinity, water levels, and temperatures. In colder climates, saltmarshes are their dominant type of habitat. Saltmarshes are shallow wetlands whose dominant vegetation is grass that can withstand high salt concentrations. They contain lots of biodiversity and act as a storm buffer zone. They act as fish nurseries and also catch sediment before it runs into the ocean. Mangroves are tropical saltmarshes that also contain boundless biodiversity. The copious decomposition that accumulates in these habitats limits the oxygen available, but it enables high rates of photosynthesis. Mangroves are nesting sites for pelicans, ducks, eagles, and vultures, and are usually the site for shrimp farms. Because estuaries are so ecologically important, the United States established the **National Estuary Program** in 1987 to identify, restore, and protect significant estuaries.

Angelina Dimitrova/Shutterstock.com

A Desert Lake

Estuaries are among the most rapidly developing areas on earth; Boston and San Francisco are built on estuaries. These urban areas tend to be severely polluted, and high levels of toxic metals are found in the ecosystem. Such metals are subject to **biomagnification.** Furthermore, the flushing water and sea spray from estuaries play a key role in water and air purification. Urban development and pollution inhibit these natural ecological functions.

Shannon Matteson/Shutterstock.com

A Mangrove

Irrigation

Approximately 75% of the freshwater supply goes to agriculture; usually employing a method called flood irrigation. Flood irrigation involves flooding or soaking crops with water. Although a simple and time-saving application, flood irrigation wastes exorbitant amounts of water and roughly 70% is lost through evaporation.

An alternative is drip irrigation, a method of irrigation that minimizes water use. Although more costly, its application (of fertilizer as well as water) involves using tubing, piping, and valves to water the roots of plants with great precision. Drip irrigation decreases soil runoff

Edmund Lowe Photography/Shutterstock.com

Flood Irrigation of Rice Fields

Jutia/Shutterstock.com

Drip Irrigation

and salinization, reduces the demand from depleting water sources such as aquifers, requires less energy, and increases crop yield.

Water Crisis

Seventy-one percent of the earth's surface is covered by water of which 97% is found in the oceans, 2% are in glaciers or ice caps, and roughly 1% is fresh water, some of which is available to us as drinking water. The majority of our drinking water supply comes from melting mountain glaciers (that turn into mountain springs), reservoirs, and aquifers. Because such a low percentage of the earth's water supply is freshwater, and also because a high percentage of water from the hydrologic cycle ends up in the oceans (where it is not available to us as freshwater), freshwater is considered a nonrenewable resource. The growing human population (expected to grow almost 50% in the next 50 years!) exhausts the already limited supply. Already, almost 1 billion people on this planet do not have access to clean drinking water, and over 3 million die each year from a water-related disease.

To solve the problem of water shortage, we will explore several solutions that would provide the growing human population with clean water.

Domestic Water Use in Gallons per Day per Person and Projected Percent population Change by 2030

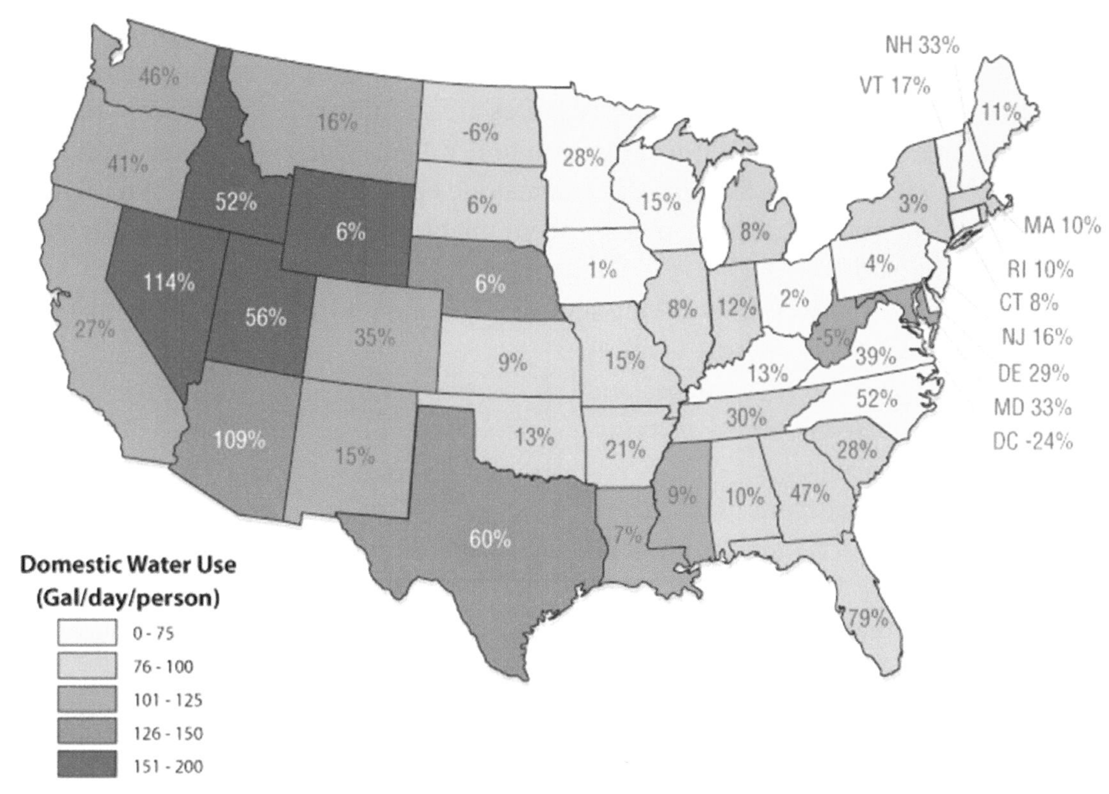

Water data from USGS, Estimated Use of Water in the United States in 2005. Table 6, Page 20; population data from U.S. Census Bureau, State Interim Population Projections by Age and Sex: 2004–2030.

Dams

Oleksandr Koretskyi/Shutterstock.com

The Glen Canyon Dam, Lake Powell, and the Colorado River

To create freshwater supply, dams, such as the Hoover Dam in Nevada, are built to stop the flow of rivers (or in this case the Colorado River) and create a body of freshwater called a reservoir. Reservoirs not only provide drinking water, but they also have multipurpose for functions such as flood control, generation of electricity, irrigation, and recreation.

Although multiuse and therefore convenient, dams create a very large land disturbance. As water is dammed up, reservoirs are created on what was often once dry land, forcing people to relocate (if they inhabited the area). This can be a massive undertaking for cities and towns and the residents themselves. Dams also have many other negative ecological impacts such as inhibiting fish spawning and the acceleration of decomposition rates in shallow water, thus increasing the amount of CO_2 released into the atmosphere.

Water Purification

Municipal water must go through a treatment process which purifies water to ensure it can be reintroduced to the environment. **Potable** water is purified water that is safe to drink. One mechanism used to create potable water is **reverse osmosis**, the process in which water is pushed through a synthetic membrane whose pores are so miniscule that all impurities and contaminants are filtered out. Then it is exposed to UV light which is lethal to most fungal spores, bacteria, viruses, protozoa, algae, and nematode eggs. UV light causes mutations and changes in DNA of microorganisms, making them unable to reproduce. Finally, it is treated with hydrogen peroxide.

Another mechanism involves purifying sewage waste and returning it to a state that is safe for the environment. This process involves primary, secondary, and tertiary stages and can be done locally (septic tanks) or transported to municipal plants.

irabel8/Shutterstock.com

A Sewage Treatment Plant

Primary sewage treatment uses a screen to filter out and extract water from solid waste. The solid waste that remains is called **sludge** and is eventually dried and used as fertilizer.

Secondary treatment involves adding bacteria to the filtered water in order to break down any remaining organic matter.

Tertiary treatment involves using UV light to kill off any remaining bacteria, viruses, and other microorganisms. Finally, is treated with chlorine at which point it becomes potable.

Kekyalyaynen/Shutterstock.com

A Desalinization Plant

Another purification process involves water being treated by **desalinization plants** whose purpose is to remove salt from ocean water. This can be done by either of two mechanisms. The first is to distill the saltwater, and since salt has a different boiling point than water, salt will separate during distillation. Water vapor is collected and cooled to a liquid state and is then used as drinking water. The other desalinization alternative is **microfiltration**, a process in which water is pushed through a membrane-like screen which mechanically separates water from salt.

Review Questions

1. What are some unique adaptations of camels that enable them to withstand long periods without water?

2. What are the different zones of the ocean, from the euphotic zone downward? What unique adaptations of aquatic life would you find in each zone?

3. Why are corals considered a foundation species?

4. What unique adaptations would you find in wildlife in the intertidal zones?

5. What are differences between eutrophic and oligotrophic lakes?

6. Which is more sustainable, drip or flood irrigation? Why or why not?

7. Discuss the different stages involved in the treatment of sewage.

Discussion Questions for Class

A. Should we pay more for drinking water? Would that help the water crisis?

B. How have the root causes of environmental problems influenced availability of water?

C. Discuss how human activity has polluted the different aquatic ecosystems.

D. Discuss the pros and cons of fish farms vs. naturally caught fish.

E. Discuss the benefits of rainwater harvesting rebates and sustainable landscape rebates for water efficiency.

F. Discuss the benefits of high-efficiency clothes washers, and sprinkler rotary nozzles.

G. An obstacle to desalinization is how to store excess salt. Dumped into the oceans, it creates briny water that is an ecological disaster. Furthermore, distillation uses a lot of energy and thus has a large carbon footprint. How do we deal with the obstacles of obtaining freshwater?

Your Personal Sustainability Journal

Educate yourself on origin of the seafood that you eat. When possible, eat plant-based alternatives.

Donate or volunteer for non-profits such as the Coastal Conservancy (coastalconservancy.org) or Ghost Fishing (www.ghostfishing.org).

Use grey water—recycle water from washing dishes or the bath to wash the car.

Turn the shower off when you are shampooing or rinsing your hair

Take timed, 5-minute showers.

Take long baths instead of showers.

Stop buying bottled water—use a filter for the tap.

Design your own rainwater harvesting container. Leave it outside during the rainy season and use the grey water to wash your car or water your lawn.

Install water-saving devices on your faucets and shower heads, and fix any leaks in toilets, faucets, or fixtures.

Add your own suggestions here.

Endnotes

i. Thomas J. Webb, Beth L. Mindel. Global Patterns of Extinction Risk in Marine and Non-marine Systems. *Current Biology*, 2015; DOI: 10.1016/j.cub.2014.12.023

ii. Raina, Tapiolas, Foret, et al. "DMSP biosynthesis by an animal and its role in coral thermal stress response" Nature, 2013

iii. http://www.fao.org/docrep/019/i3640e/i3640e.pdf

iv. The National Oceanic and Atmospheric Administration. April 9, 2014.

v. Investigación y Desarrollo. "Over demanding market affects fisheries more than climate change." ScienceDaily, 7 March 2014. <www.sciencedaily.com/releases/2014/03/140307165915.htm>.

vi. Please see www.econyl.com

vii. Referenced by Heather Hamza through Ghost Fishing USA

Chapter 10

Earth's Biomes

DID YOU KNOW?

Certain pesticides, toxic chemicals, and heavy metals that are manufactured in warm climates can heat and evaporate into the atmosphere. With decreasing atmospheric temperatures, they cool and deposit in regions with colder climates—places such as the Arctic. This is called the **grasshopper effect** or, more technically, **global distillation**. This might explain why DDT (dichlorodiphenyltrichloroethane) and mercury are found in abundance in the otherwise pristine and untouched Arctic tundra. Here, they enter the ecosystem where they **bioaccumulate** until they are found in substantial quantities in caribou and polar bears. This unfortunate phenomenon is a harsh reminder that pollution and environmental issues are a *worldwide* problem.

Climate Influences Biomes

As a result of the curved surface of earth, prevailing winds in the northern hemisphere moving from high to low pressure seemingly deflect right (clockwise) while prevailing winds in the southern hemisphere moving from high to low pressure seemingly deflect left (counterclockwise). We call this the **Coriolis effect**. The Coriolis effect is believed to be responsible for large-scale weather patterns, but given its complexity and the many variables involved, scientists are not quite sure how this is so.

Because the earth rotates on an axis, the sun heats the earth's surface unequally, causing much variation in weather patterns. **Weather** is a result of this unequal heating and is defined as *short-term* fluctuations in such variables as temperature, humidity, precipitation, and cloud cover. We experience the variation of these *temporary* conditions on a daily basis.

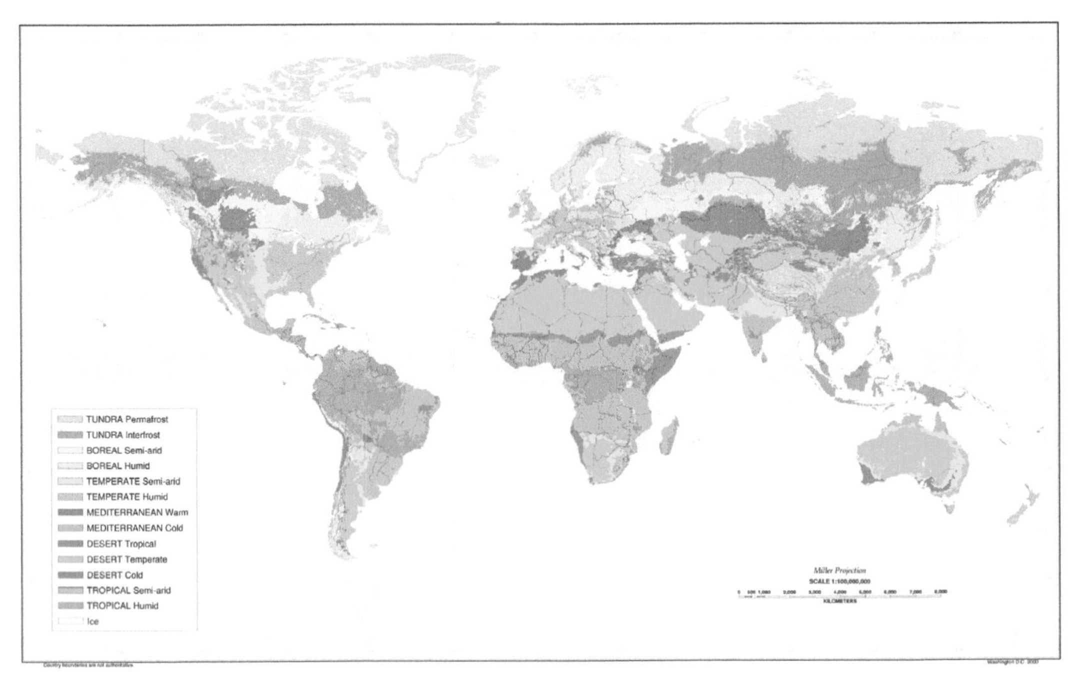

Major Climates

Source: Soil climate map, USDA-NRCS, Soil Science Division, World Soil Resources, Washington D.C.

The National Oceanic and Atmospheric Administration (**NOAA**) is the Federal agency that researches weather patterns, changes in our oceans, and in our climate

Climate is defined as *long-term* trends in weather patterns of a large given land area. Thus, long-term weather patterns directly determine climate. There is an observable correlation between latitude and climate. Because the earth spins on its axis, the equator gets the most net sunlight per year. The further distance on ecosystem is from the earth's equator, the colder its climate and average daily temperature. There is also an observable trend between altitude and climate—the higher the altitude, the colder the climate. That may explain why mountain peaks often resemble the Arctic tundra!

Biomes are observable trends of vegetation in terrestrial ecosystems that are directly correlated to climate. They are the highest level of the ecological system. Biomes are characterized by their dominant plant life which, in turn, is adapted to the climate of a given biome. For example, Southern California experiences a Mediterranean climate, marked by long, hot, dry summers and cool, wet winters. A dominant biome found in Southern California is chaparral—or Mediterranean woodland. Chaparral is comprised of vegetation that is adapted to the Mediterranean climate and therefore can endure long, hot summers and cool, wet winters (some adaptations include reduced waxy leaves—designed to reduce water loss). Other parts of the world that experience Mediterranean climates include central Chile, Italy, Greece, and Southern Australia. Each of these countries has the Mediterranean woodland biome. Although their vegetation does not share a recent evolutionary history with the chaparral found in California, they look very similar to the chaparral of California. The similarities are attributed to the fact that they have also individually adapted to the Mediterranean climate (this is an example of **convergent evolution**).

Ecotones are found where two distinct biomes overlap. They are a hotbed of biodiversity because a variety of plant and wildlife, adapted to a variety of environmental factors and two separate biomes, are found here.

Tundra

The tundra is a biome found in either high latitudes (the Arctic tundra) or high altitudes (the Alpine tundra on mountain tops). The Arctic summers are very short, lasting only a couple of months, but summers experience almost constant daylight. Precipitation varies from 200 to 600 mm yearly, and precipitation exceeds evaporation. This is the coldest biome and is distinguished from all others by the presence of **permafrost**.

Because the ground temperature is cold, decomposition rates are low creating this permafrost, a layer of frozen ground littered with layers of decaying or dead leaves and **detritus** (dead organic matter). The tundra is subject to **solifluction**—when frozen, water-saturated soil slowly moves down-slope. Dominated by grass, moss, sedges, lichen, and small shrubbery, the vegetation here is adapted to maximize heat absorption (they are mostly oriented toward the ground to absorb heat from the earth). Much of the vegetation has dark pigmentation to increase heat absorption. In addition, leaves orient themselves perpendicular to the sun to maximize sun exposure. Wildlife found in this biome includes polar bears, the Arctic fox, wolves, musk, ox, reindeer, ground squirrels, weasels, snowy owls, lemmings, and caribou.

Kekyalyaynen/Shutterstock.com

Permafrost

Sergey Krasnoshchokov/Shutterstock.com

Caribou in Russia

The tundra is being adversely affected by the increase in global temperatures. Due to these temperature changes, many animals inhabiting most southern biomes are migrating north to the Arctic tundra. The permafrost is melting, compromising the foundation of roads and infrastructure built on the frozen ground. Carbon dioxide trapped in the permafrost is released into the atmosphere as the permafrost melts, further impacting global temperatures. Furthermore, as the permafrost melts, the dark ground cover beneath it is exposed. Dark colors absorb more wavelengths of light, which also increases the overall temperature of the Arctic biome.

The Arctic has, until recently, been considered pristine, with little disturbance from human activity. However, recent years have seen more oil extraction, facilitating more human activity and disturbance. Furthermore, the tundra is vulnerable to the grasshopper effect, as pesticides and chemicals from distant populations accumulate in the Arctic. In addition to DDT and mercury (see the paragraph above), cesium-137, a radioactive element released in the devastating Chernobyl power plant disaster has been found in the tundra of Norway.

Taiga

The taiga (or the boreal forest or coniferous forest) is a biome found in the northern hemisphere in such places as Canada, Scandinavia, and Russia. A typical pattern of the taiga is picturesque freshwater lakes surrounded by dense rows of evergreens. The freshwater lake is what remains of a glacier that receded thousands of years ago.

Mark Herreid/Shutterstock.com

The taiga is found to the north and is marked by long severe winters (that last over 6 months) with little precipitation—about 20–60 cm a year. Temperature ranges from 70°F in the summer to 30°F in the winter.

The vegetation found here includes spruce, fir, and pine trees as dominant vegetation, but one also finds ferns, grass, and moss as the ecosystem producers. Since the ground is frozen for prolonged periods of time, the conifers (cone-producing trees like pine and spruce) are adapted to prevent water loss. Their leaves are thin, waxy, needle-like, and have minimal surface areas to prevent water loss through transpiration. The forest floors are nutrient poor and acidic. Pine needles drop to the ground, blanketing the forest floor, causing a decrease in soil pH. Acidity and cold temperature slows decomposition rate, causing a deficiency of available soil nutrients. Because soil quality is poor, and the long winters do not make for a suitable growing season, the Taiga is not suited for agriculture. However, they are a leading provider of lumber, paper, and other forest products. Drilling for oil, gas, and mining also occur in the Taiga.

Types of wildlife found here include moose, caribou, small rodents, birds, reindeer, bison, black and brown bears, wolves, lynx, and hares.

The taiga is being degraded by increased hunting and trapping of these animals and by pollution and disturbances associated with mining, logging, and oil drilling.

Deciduous Forest

Deciduous forest biomes are comprised of broad-leaved hickory, maple, beech, and oak trees, which lose their leaves annually. One also finds ferns and flowers as common producers. These cool and moist forests are home to some of the earth's largest organisms. Although

Petr Baumann/Shutterstock.com

winters are cold, temperatures are not as extreme as they are in the taiga. Its growing season (summer) is at least 4 months long and provides ample precipitation; annual rainfall reaches anywhere from 65 to 300 cm per year, with a temperate climate in which temperatures range from −30°C in winter to 30°C in summer. Deciduous trees lose their leaves in winter months to prevent water loss when the ground is frozen (and precipitation often comes down as snow). Some wildlife found here are deer, rodents, skunks, bear, fox, cougars, and wolves.

Because these forest have an abundance of leaf litter, long growing seasons, and moderate temperatures, its soils experience rapid decomposition in warmer months, making soils rich and fertile with dead organic matter. Forest floors are often dominated by moss and mushrooms as well as microscopic bacteria that contribute to decomposition of the dead wood and other organic matter, thereby recycling nutrients.

Deciduous forests are being degraded as a result of human population growth and development. Many major cities such as Boston, DC, and London are built on what were once deciduous forests. They were also among the first biomes converted to agricultural land given the good soil quality and fertility. Urbanization, pollution, over-grazing, and deforestation have all disturbed this soil-rich biome.

Grasslands

Grassland biomes are found in the interior of continents where the annual rainfall is too little to sustain forest growth, but too abundant for a desert biome. They are characterized by fertile soils and herbaceous (nonwoody) vegetation.

Grasslands are dominated by either long or short grasses. Longer grasses thrive with more moisture and precipitation.

Papa Bravo/Shutterstock.com

Zebra Running on the Savannah Grassland

Tropical grasslands, such as the Savannah, found within 20° of the equator, experience both dry and wet seasons. Soil here is dense and does not retain water, making it difficult for trees to grow. Dry seasons occur in the summer and are often accompanied by fire which release nutrients from the soil that are imperative to plant growth, increased fertility, and resprouting of grasses. The African Savannah is home to many animals such as giraffes, zebras, lions, and elephants.

Savannah grasslands experience frequent fire which often kills its vegetation. The grass survives and resprouts. In this biome as well as many biomes, fire can have many ecological benefits.

> **Surface fires** control pest and pathogen populations, release nutrients from the soil, enable vegetation to crown sprout, and nuts to crack open, releasing seeds (sequoias are an example). Periodic surface fires are also important for the Carbon cycle. Fire departments often utilize prescribed (or controlled) fires to mimic earth's natural fire ecology and gain its ecological benefits. **Crown fires** on the other hand are damaging. They burn hot, kill trees and wildlife, and cause soil erosion.

Temperate grasslands are the largest biome in North America and are even more expansive in Europe and Asia. The dominant vegetation is goldenrod, sunflowers, and clover. In temperate grasslands (such as the prairie grasslands of Montana), annual precipitation reaches 30–100 cm per year, with summer months receiving the most precipitation. Winters are cold, and summers are long and hot. Soils are very rich with nutrients and dead organic matter. Temperate grasslands are home to coyotes, bobcats, grey wolves, bison, and hundreds of bird species.

Prairie Grassland of Montana

Because their soil is rich and fertile, a large portion of prairie grasslands have been plowed over and farmed, causing them to lose up to 40% of their nutrients. Grasslands are also being degraded by overgrazing.

Desert

Desert biomes are often found ~30° north and south of the equator, just north and south of the subtropics. Here, tropical air descends, drying out and forming deserts. They are characterized by drought, flash-flooding, heat, bitter cold, and **lithosols**.

Sahara Desert

Deserts are expanses of land where *evaporation exceeds precipitation*. Their soil holds meager nutrients but often contain an abundance of stone, mineral, and salt. The vegetation that grows here are adapted to minimize exposure to sun and water loss. The leaves are usually small and waxy, and shed during dry periods to prevent water loss. Such examples include cacti, sagebrush, and yuccas. In the summer months, desert animals are usually active at dusk or night to avoid sun exposure.

Galyna Andrushko/Shutterstock.com

Gobi Desert

Desert biomes are characterized by *hot desert, temperate desert*, or *cold desert*.

Hot deserts, such as the Sahara, are hot most of the year with temperatures reaching over 45°C. Temperate deserts, such as the Mojave in California, have almost no water vapor in the atmosphere, and the temperature can get hot during the day and very cold at night, yielding extreme variations daily. And finally the cold deserts, such as the Gobi, have long and cold winters.

The producers of desert biomes include cacti, ocotillo (desert coral), brush, and creosote bushes. For wildlife, the desert is home to lizards, snakes, scorpions, hawks, kangaroo rats, rodents, and foxes.

In all biomes, disturbance by humans, overexploitation, human settlements, mining, overgrazing, agriculture, etc., has caused desertification of the soil, increasing the number of desert biomes worldwide. Deserts are fragile ecosystems since their soils hold few nutrients. Disturbed desert ecosystems take many years to recover.

Tropical Rainforests

Tropical rainforests are found near the equator and experience almost daily rainfall. Comprising only about 6% of the earth's land mass, they contain an estimated 50% of the world's biodiversity on land!

Microstock Man/Shutterstock.com

Tropical rainforests experience little temperature fluctuation from month to month with the average daily temperature being 25°C. Annual rainfall can reach almost 400 cm per year. Rainforests are among the oldest forests found on earth. The tops of rainforest trees collectively create a dense impenetrable canopy, allowing little sunlight to reach the forest floor. As a result, trees and other vegetation must compete for sunlight. Many **epiphytes** live on rainforest trees, using the host tree as a facilitator in their vertical growth to access to sunlight. This relationship between the rainforest tree and the epiphyte is a type of **commensalism**.

The soil of the rainforests is nutrient poor with the constant rainfall leaching nutrients from the soil. Rapid decomposition from bacteria, ants, and fungi keeps soil nutrient content low as the nutrients are quickly absorbed by tree roots.

Trees (evergreen angiosperms) dominate this landscape and can grow up to 50 m in height with extensive canopies which themselves are homes to as many as 500 species—including plants, salamanders, birds (such as brilliantly colored parrots), snakes, frogs, and monkeys. Due to the extensive canopies, very little sunlight reaches the forest floor, so vegetation is not dense at ground level.

Deforestation, logging, mineral extraction, and agriculture destroy an estimated 100,000 square kilometers of rainforests per year.

In some areas, much of the Rainforest trees are cut down to make grazing land for cattle and to grow soy in order to feed them. Palm oil plantations are also a major driver of Rainforest destruction.

Temperate Rainforests

This cool, damp ecosystem stretches along the United States West Coast from southern Oregon to southeast Alaska. It also grows in Southern Australia, New Zealand, and Chile. Some of the common forest trees found in this ecosystem are very large evergreen angiosperms: Western Hemlocks, Douglas firs, Red Alder, and Bigleaf Maples that provide us with most of our lumber and wood pulp. They can grow over hundreds of years old, 60 feet in circumference, and 250 feet in height. They typically have epiphytes growing in the canopies; and moss, ferns, lichen and Huckleberry in the understory. These temperate rainforest receive plenty of annual rainfall—roughly 12–14 feet per year. Its mild temperatures rarely reach above 80°F or go below freezing. Temperate rainforests are typically found near the coast and the condensation from the marine fog add to the overall precipitation and dampness. Lots of dead and decaying wood, leaves and needles are found on the forest floor. This makes for new habitats for insects, moss, amphibians fungi and small mammals. Deer, elk, wood rats, and squirrels are common inhabitants.

A Temperate Rainforest

These old-growth forests have been over harvested for logging—to provide us with our pulpwood and lumber. At least in the United States, only a small percentage of this old growth remains. Most have been replaced with tree farms, which lack the biodiversity of the primary growth.

Chaparral/Mediterranean Woodland

This biome, found in Mediterranean climates, is characterized by long, hot summers and cool, wet winters.

Zack Frank/Shutterstock.com

This biome has high species richness. Some of the plants have flammable essential oils that promote fires which, in turn, promote crown sprouting and fertility. Its soil is not considered nutrient-rich due to slow decomposition rates.

Chaparral vegetation is evergreen and typically has small, reduced waxy leaves to prevent and conserve water. Here (at least in California), you find sages, toyon, laurel, sumac, walnut, and prickly pear.

Much of the wildlife here are nocturnal; they burrow during the day to protect themselves from fire, escape the heat, and prevent water loss. An adaptation of some wildlife is to secrete concentrated urine in order to remain hydrated. Its wildlife includes hundreds of species of migratory birds, wood rats, lizards, deer, sheep, and in the Mediterranean woodland of Australia ... kangaroos!

Chaparral is not without human disturbance. Many people prefer a mild Mediterranean climate to extreme long winters, and move or retire to this biome. Thus, high human density and development are common here. Humans clear the vegetation for agriculture and grazing. They also set fires (prescribed fires) to control pests and prevent more intense crown fires.

Climate Change Causes Biomes to Shift

Many terrestrial areas, or biomes, are at risk of shifting to a new type of natural vegetation as a result of changes in climactic patters. These changes in climate are associated with anthropogenic greenhouse gases.

Such climactic factors as wind, soil moisture, temperature, humidity, and precipitation influence or limit the geographic ranges of most animal and plant species. Any shift in the above variables may have a profound effect on plant and animal ranges and on biomes altogether. Why is this so?

Because Biomes are the highest level of an ecological system, a shift in these biomes can create a profound force, or ecological shift. When biomes change, wildlife or plants must adapt accordingly to the new variables. Wildlife or plants that cannot adapt to change may go extinct or disappear locally. For examples, Climate change, increasing temperatures and other factors, have shifted vegetation upslope, toward poles or toward the equator. It is predicted that the taiga, boreal forests (evergreen coniferous forests), which are adapted to colder temperatures, will move toward the poles. Research suggests that alpine plants have already shifted their range.[i] But the survival of some plant species may be at risk if climate changes faster than trees can seed, grow, and settle into new areas.[ii]

Many animals respond to changes in climate faster than plants. It is predicted that we may see many animals migrate north. Species with a narrow range of tolerance for temperature may seek out higher altitudes and latitudes as global temperatures get warmer Range shifts have already been reported in animal species such as birds,[iii] butterflies,[iv] and mosquitoes.[v]

Some animals are reportedly changing their diet with climate change and migration.[vi]

It is unpredictable how this will play out and how biomes will redistribute, but scientists are certain that there will be a major shift in most biomes as their plant species and animal species redistribute their ranges.[vii]

Review Questions

1. Compare soil quality of all the biomes. Which biome has nutrient-rich soil? Which has nutrient-poor soil? Which experience high decomposition rates? Which experience low decomposition rates? Why or why not?

2. Why are desert biomes associated with flash flooding?

3. Explain why cesium-137 has been observed in the Arctic tundra of Norway. Where did it originate and why is it found in the tundra biome?

4. Discuss some ecological benefits of fire.

Discussion Questions for Class

A. All earth's biomes are threatened by human consumption. How have root causes of environmental problems threatened earth's biomes?

B. Global weirding, El Nino Southern oscillation is a real phenomenon. We are experiencing changes in global temperatures. How will this affect the different biomes?

C. Given the phenomenon of the grasshopper effect, do you consider pollution a local or worldwide problem? Why or why not?

Your Personal Sustainability Journal

Check on the wildlife habitat around your home or office. Note the presence of any new wildlife or growth.

Go for a hike or walk in nature of your local community. Which biome do you inhabit? Notice the adaptations of the plant life and wildlife. Think of ways to conserve their habitats.

Donate to or get involved with the Sierra Club! (www.sierraclub.org).

Add your own suggestions here.

Endnotes

i. Grabherr, G. M. Gottfried, and H. Pauli, 1994. Climate effects on mountain plants. Nature 369, 448
ii. IPCC, 1998. The Regional Impacts of Climate Change: An Assessment of Vulnerability, (Eds RT Watson, MC Zinyowera, RH Moss), Cambridge University Press, Cambridge, UK. (Appendix C)
iii. Thomas, C.D. and Lennon, J.J., 1999. Birds extend their ranges northwards. Nature 399: 213.
iv. Paremsan, C., et al. 1999. Poleward shifts in geographical ranges of butterfly species associated with regional warming. Nature 399, 579–583.
v. Epstein, P., H. Diaz, S Elias, G Grabherr, N. Graham, W. Martens, E.M. Thompson, and J. Susskind 1998. Biological and physical signs of climate change: focus on mosquito borne diseases. Bulletin of the American Meteorological Society 79, 409–417.
vi. James Buckley, Jon R. Bridle. Loss of adaptive variation during evolutionary responses to climate change. *Ecology Letters*, 2014; DOI: 10.1111/ele.12340
vii. IPCC, 1998. The Regional Impacts of Climate Change: An Assessment of Vulnerability, (Eds RT Watson, MC Zinyowera, RH Moss), Cambridge University Press, Cambridge, UK.

Chapter 11

Bioengineering, GMOs, and Agriculture

DID YOU KNOW?

Monsanto—a Missouri-based bioengineering company specializing in GMO (genetically modified organism) agriculture—developed a genetically modified corn that is resistant to pesticides and herbicides. Their GMO pollen dispersed and hybridized with local family-owned farms (unbeknownst to the family farmers) whose corn was grown organically. These family farmers had saved and replanted seeds the following year, a common agricultural practice.

Asserting that they were protecting their work and innovation, Monsanto sued the local farmers claiming patent infringement. As a blow to small-scale family farmers, Monsanto won the lawsuit. Monsanto's GMOs have been identified as an accidental introduction to many small farm operational seed supplies, which can sometimes result in inviable hybrid seeds.

According to the Public Patent Foundation, family farmers and organizations with organic interests have preemptively sued Monsanto to protect themselves and their interests from Monsanto lawsuits in the future. Their pending lawsuit has set precedence to protect farmers and their organic crops from obsolescence through hybridization with GMOs.

Manipulating Nature

Biotechnology is the science of manipulating living organisms for human needs and benefits. Wine-making and cheese-making are common examples of biotechnology. **Genetic engineering** is the process of *altering the DNA* of a living organism to create products of human interest. **GMOs** are the product of genetic engineering. Reasons for altering DNA include introducing a beneficial trait into an organism or deleting an undesirable trait from an organism.

Gokce Gurellier/Shutterstock.com

Some desirable results of genetic engineering include the slow ripening of fruit, higher yield of crops, the inclusion of nutrients not otherwise found in regional crops and staple foods, and the creation of crops resistant to drought, salinity, pesticides, and other conditions that may otherwise decrease agricultural production.

One example of nutritional fortification of foods can be seen with beta-carotene. Deficiency in dietary beta-carotene can lead to blindness. In many developing nations, the rate of blindness is particularly high and has been increasing due to malnourishment. The use of biotechnology has allowed the introduction of beta-carotene.

It is interesting to note that almost 75% of the food in U.S. supermarkets is genetically modified. According to the International Crops Research Institute for the Semi-Arid Tropics (ICRISAT), over 70% of the U.S. crops are transgenic (formed from two or more DNA sources), a much higher percentage than any other country in the world. Some of the most common U.S. GMOs are cotton, soy, and corn, apples, alfalfa, potatoes, and squash.

cholder/Shutterstock.com

Rows of Cotton

Tish1/Shutterstock.com

Rows of Corn

GMOs are also used to create new drugs and to harvest proteins and hormones such as insulin and growth hormone (GH). Bioengineering also performs bioremediation through microbes (bacterial DNA makes proteins that degrade waste from pesticides and pollution).

How Does Genetic Engineering Work?

Bacteria cells have circular noncoding DNA called **plasmids** that act as storehouses of DNA. Plasmids are able to take DNA from the environment and incorporate it into the genome of its bacterial cell. This is a normal practice of bacteria as it confers beneficial traits such as resistance to antibiotics that may exist in their environment. The contents of the DNA plasmid can be easily manipulated in the lab for our benefit.

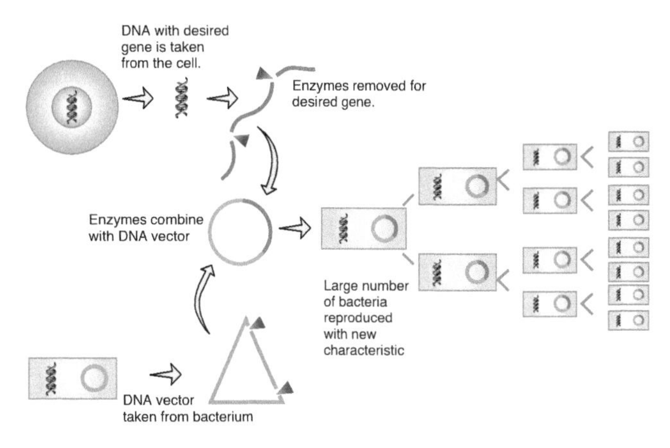

An example of what bacterial use of plasmids can do for us is exhibited in how scientists use genetic engineering to create and harvest bovine growth hormone (BGH). BGH increases the yield of milk that a cow can produce. The desired DNA sequence (that confers this increased milk production ability) is isolated from cow cells by using a **restriction enzyme**—a type of DNA scissor that can find a specific sequence of DNA and snip it from the irrelevant DNA. This targets our BGH trait and creates a DNA fragment that can be collected. The bacterial plasmid is cut with the same restriction enzyme so that the BGH DNA fragment is now complementary to the plasmid DNA (meaning the two types of DNA can now be put together). This results in what are called **sticky ends**. The DNA from the two sources is spliced together, and the result is a **transgenic** bacteria. Transgenic means that its DNA comes from two (or more) separate sources.

It is important to note that a plasmid is known as a **vector**, as it essentially carries the trait of interest (BGH) into the bacteria which, in turn, acts as a factory for the protein that will be produced from the trait of interest. The bacteria are cultivated on a growth medium and allowed to produce the protein that our DNA encodes (BGH), and the hormone is extracted and stored for later use.

Pros and Cons of Genetic Engineering

Pros

Genetic engineering can enable crops to grow larger and more colorful, become seedless, ripen slowly, increase yield, become drought, pesticide, and herbicide resistant, and can enable crops to grow on marginal to low-quality soil. In fact, bioengineering performs many of the same agricultural processes that farmers have done for thousands of years (artificial selection for desired traits), but in a much shorter period of time. This is accomplished by inserting the desired gene (that produces a desired trait) instead of taking generations to crossbreed. Traditionally, crossbreeding has been performed with closely related crops species, but bioengineering allows genes to be introduced from many different species of different organisms. In fact, most of the genes bioengineered into crops come from bacteria. Also, genetic engineering helped save Hawaii's papaya industry by helping make them resistant to the papaya ringspot virus, which threatened to wipe them out.

Cons

Criticisms or concerns of GMOs are numerous. One concern is that crops have coevolved over thousands of years with their pollinators (bees, hummingbirds, etc.) and have adapted their own ways to ensure pollination and evade pests. (In fact, spiders have killed more crop pests than any synthetic pesticide.) This can lead to an imbalance or collapse of many populations of organisms in a particular environment by disrupting the food chain. Consider that herbicide-resistant crops often grow more aggressively, requiring more pesticide and herbicide. In fact, the crop ultimately becomes a weed during crop rotation. Plants and crops are difficult, if not impossible, to contain, and GMOs may hybridize with natural, nongenetically modified organisms and disrupt their natural ecological role. Therefore, bioengineering crops may disrupt thousands of years of coevolution and have vast ecological consequences. Furthermore, engineering crops can cause a reduction of genetic diversity of crops. This can make crops more vulnerable to parasites—and situations similar to the Irish Potato Famine for another con: it may make food crops susceptible to antibiotic resistance..

Still another criticism is foul play. A terminator gene has been introduced into some GMOs, making these crops sterile by keeping them from proliferating for a second growing season. An expressed concern is that this feature is fueled by monetary incentive so farmers have to keep buying new seeds each year. This is contrary to the normal practice of families of farmers who may have developed their own family-owned seeds for use. Cross-pollination from a GMO field to a non-GMO is virtually impossible to contain, but can be detected by proprietary seed-producing companies. This situation has in the past caused small business farmers to be sued and ultimately lose property for having their family-based seeds contaminated with a seed of company's genetics.

To address the "pro" that GMOs would be ideal for drought-prone areas: Plants respond in complex and variable ways to drought. This is reflected in the fact that there are multiple genes that code for drought resistance and are expressed at different times and during various stages of the life cycle. This is an adaptive advantage, since droughts vary in how long they last and how severe they are–and at what stages of growth they affect plants. But genetic engineering can only manipulate a few genes at a time, so any one engineered gene may be more successful than others at any given time. But it is difficult to anticipate or predict.

Furthermore, genes that code for drought resistance may have other (sometimes undesirable) effects on crop growth. This is called **pleiotropy**. This interconnectedness and complexity suggests that drought resistance correlates with other aspects of plant growth. Manipulating these genes may carry out the undesirable genetic expression.

Yet another concern is the fact that GMO use in the developing nations has surpassed their use in the developed world. This is harmful to farmers in developing nations who have become increasingly dependent on the United States for their food and seed supply. Farmers in developing nations have lost the knowledge to subsistence farm (farm to survive) and so have become more dependent in growing their food supply.

ivancreative/Shutterstock.com

Phent/Shutterstock.com

The final concern of GMOs is the unknown. We still do not know the full extent of side effects from bioengineering crops, or if the benefits outweigh the costs. Consider that a gene from the bacteria *Bacillus thuringiensis* serves as a natural pesticide for corn and potatoes. It produces a toxin BT that controls caterpillar populations (a common agricultural pest). It has been discovered that the toxin was found in high traces in the crops that people were consuming. Nobody knows the full health side effects of such a toxin in our bodies.

More potential risks include rearranging the plant genome, inadvertently introducing mutations, and development of cancers and viruses—all of which present unknown risks and obvious threats to humans.

Agriculture

Industrial Agriculture

Industrial farming typically involves the use of a **monoculture**, a practice in which one crop is grown at a time in a given area. A uniform chemical herbicide and pesticide (typically synthetic) is used in one fell swoop for monocultures. This yields a high number of crops and costs the least amount of time and money. Industrial agriculture has saved many lives by providing large quantities and varieties of crops at affordable prices. By using synthetic fertilizers, farmers are also able to grow crops in marginal soils; by using pesticides and herbicides, farmers have been able to avoid possible pest outbreaks and famine.

Although efficient, industrial agriculture can be unsustainable and environmentally detrimental, exhausting land resources. It is estimated that we have lost one-third of our topsoil in the last 100 years as a result of our farming methods—particularly monoculture. Because of the resulting poor soil quality, industrial farming relies more and more heavily on nitrogen-rich fertilizers. The nitrogen ultimately leaches into our lakes and streams, resulting in fish kills and **dead zones** (places of low biodiversity). The nitrogen also reacts to form the secondary

Case Study

Why are honeybee populations declining?

This is a complicated question with many layers, and there may not be a single explanation. But according to Dr. Marla Spivak, an entomologist from the University of Minnesota, monoculture and industrialized farming may be partly to blame. Monoculture, along with a combination of other factors, affects bees by limiting their access to food.

Take into consideration that many honeybees are infected by a widespread parasite that circulates a virus throughout their body, weakening them. This may inhibit their mobility and their ability to effectively visit a food source. To compound the issue, their food source may be too distant for them to access. Why would this be? Before industrialized farming in the United States, farmers grew vegetation such as alfalfa and clover as a cover crop—a good food source for bees. They also enriched the soil with nitrogen. With the onset of industrialized farming, farmers stopped growing these cover crops (moving toward synthetic fertilizers) therefore taking away a bee food source. To add yet another variable to the issue, chemicals found in many synthetic pesticides may disorient honeybees, even further inhibiting them from reaching their food source. Therefore, a combination of parasites, pesticides, and industrialized farming may be responsible for the decline in honeybees.[iii,iv]

Daniel Prudek/Shutterstock.com

pollutant nitrous oxide, a greenhouse gas. In addition, many of the pesticides, herbicides, and insecticides used are controversial. Many forests with rich soil and lots of biodiversity are cut down to make room and to be converted to agricultural land (see the case study on palm oil below). Furthermore, agriculture accounts for 14% of global greenhouse gas emissions.[ii]

Pest Management

Since the agriculture revolution 12,000 years ago, humans have been using and experimenting with different pesticides to control pests—from sulfur to smoke, from natural oils to ash, from lime to arsenic—and different civilizations have experimented with and employed many varied treatments for pest control.

So what exactly is a pest? A pest is any organism, such as weed, insect, fungus, bird, animal, etc., that does damage to desirable plants, trees, crops, or overall ecosystem health. It may compromise animal or human health, or just be a nuisance.

In the following paragraph, we will examine common (and controversial) types of pesticides used in industrial agriculture today.

Inorganic: Inorganic pesticides are highly toxic substances that typically have high persistence in the environment. They are controversial because inorganic pesticides are often neurotoxins. Examples include arsenic, mercury, and copper.

Organic: Organic pesticides are natural compounds extracted from plants that happen to be toxic to insects. These include phenols, oils from conifers, and nicotine from the tobacco plant.

Fumigants: Fumigants are small molecules that are small enough to penetrate and sterilize the soil and typically produce a vapor or gas. They are also used to control the decay and prevent insect infestations of stored grain, and are also used to control fungal growth in the soil.

Microbial/Biological: These are living cells or organisms, or parts derived from them that naturally deter pests. Examples include fungus, virus, protozoa, ladybugs, wasps, and the BT toxin derived from bacteria (discussed above). They are designated to target certain pests. Some fungi, for example, target and kill specific insects.

Chlorinated hydrocarbons: The chemicals have high persistence and very toxic to organisms, they can bioaccumulate in adipose tissue. Examples include DDT and atrazine.

Organophosphates: These are the most commonly used synthetic pesticide today. They attack the nervous system of some animals by disrupting acetylcholine, a neurotransmitter. Organophosphates are highly toxic and short lived (they do not have high persistence). One example is roundup.

Unfortunately through natural selection, many pests have become genetically resistant to traditional pesticides that once effectively controlled their population. In addition, although effective at saving crops and therefore lives, some pesticides can become controversial due to their toxicity. They also compromise overall ecosystem health (DDT for example).

One way to control pest populations while addressing these above concerns is a strategy called **Integrative Pest Management (IPM)**. The goal of IPM is to maximize pest control by employing many various integrative methods, which effectively minimize the use of pesticides, and adjusting the use of each method accordingly. The decisions regarding which method to use is based on assessing the environment and using common sense and sound judgment to determine the best solution. For maximum effectiveness and to determine the best methodology, it is necessary to know the ecosystem and the pests involved. Monitoring the site is also imperative for long-term effectiveness. Let us examine some of the integrative methods below.

Biological or natural controls for pest populations is the strategy that mimics nature by introducing the pests' natural population controls, such as parasites, competitors, disease, inclement weather patterns, and **parasitoids** (insects such as wasps and mites that use other insects to complete their life cycle, ultimately killing them). It employs the use of pheromones to disrupt their life cycle, growing healthy crops that can resist attacks. Another method is *pesticide or chemical control.* Using alternative methods, such as fungicides, herbicides, and insecticides, pest population can be reduced. *Cultural control* employs crop rotation, agroforestry, strip farming, and other agricultural practices that are used to confuse pests on where its food is found. (See below paragraph on sustainable agriculture for more detail.) And finally, *mechanical* control involves dealing with the pest directly. This involves screens for keeping pests out, or trapping rodents.

Organic Farming

When you are at the supermarket, you may see stickers that say "certified organic." The USDA (United States Department of Agriculture) implanted the standards for "certified organic" as recently as 2002. Since that time, the organic market and consumer demand for it has increased steadily and profoundly. Testament to this is the amount of organic food available at supermarkets, co-ops, and the increased popularity of farmer's markets.

Organic farming typically involves farming that is least invasive to agricultural land, relies on renewable resources, and is least invasive to soil. These standards are met at every stage of production, from the fields to the market place. For instance, **polyculture** (growing more than one crop at a time in a given area) is practiced instead of monoculture. It also involves planting cover crops, such as alfalfa, clover, pea plants, and other legumes that put nitrogen into the soil and help maintain soil nutrients. Alfalfa may also reintroduce bee populations, as it is a food source. Organic standards apply to livestock as well. Organic meat has no hormones or antibiotics added to it.

We examined the importance of soil in Chapter 7, but let us revisit some of the key points and keep them in mind as we read about sustainable agriculture: there are 15 soil nutrients required from crop growth and shortage of any one of these can compromise crop yield. It can take 1000 years to create 1 centimeter of soil, and by the year 2050, agriculture yield must increase by 60% globally. If soil was managed more sustainably, it could yield 58% more food.[v] So let us examine more ways in which soil management and agriculture can be more sustainable:

SOIL LAYERS

O-horizon Humus
A-horizon Topsoil
B-horizon Subsoil
C-horizon Weathered rock fragments
D-horizon Bedrock

Designua/Shutterstock.com

Soil: A Mixture of Living and Nonliving Material

Crop-rotation and **no-till cultivation** are also implemented in organic farming. Crop rotation involves growing different crops each year, which maintains soil fertility, and also enables crops to evade pests naturally. This prevents pests from lingering after finding their food source. No-till cultivation eliminates the use of tractors and is less invasive to the soil than till cultivation. The result is reduced water evaporation and soil erosion. Another common application is using green manure as a fertilizer.

Natural pesticides and herbicides are used in organic farming, and the practice does not include GMOs. U.S. crops have to be only 95% organic to be considered "organic" by the FDA, and 70% organic to say it has been made with "organic ingredients." According to the *World of Organic Agriculture*, as of 2010, 35 million hectares of farmland are organic worldwide, with strongest growth being reported in Latin America and Europe.[vi]

Tran Van Thai/Shutterstock.com

Terraced Fields in Vietnam

Alternatives to conventional pesticides and herbicides include planting marigolds (a natural herbicide), scalding hot water to kill off pests, polyculture, planting mulch to control weeds, crop rotation to confuse pests and disrupt their cycles, mechanical control (weeding and physical barriers), pheromone and hormone disruptors, introducing insectaries or natural pest predators, and introducing naturally occurring microorganisms. Trap crops such as bok choy are often grown to steer pests away from food crops. The following farming techniques are designed to maximize agriculture use on land that would otherwise not be functional.

Terrace farming is typically done on steep slopes, cutting steps into the slopes in order to maintain water and to keep the ground covered with vegetation (in order to prevent runoff). Some of the ancient rice terraces that were carved into hills and mountains in countries, such as China, the Philippines, and Vietnam, are still in existence today.

Rosliak Oleksandr/Shutterstock.com

Contour Farming

David Aleksandrowicz/Shutterstock.com

Strip Farming

Contour farming involves planting crops to go along with the gentle rolling slopes of hills, instead of orienting them perpendicular to inclines. This reduces soil erosion and maximizes surface area for use.

Strip farming (or alley cropping) involves growing crops in rows to keep pests confused and to maintain soil fertility. Rows of crops are also typically flanked by tall vegetation in order to block excess wind, thereby preventing erosion. It also provides shade which reduces evaporation, maintaining soil moisture.

Agroforesty is a diverse, highly functional and integrative approach to agriculture that combines crops with shrubs, trees, and livestock. The benefits include a very biodiverse ecosystem that may elude pests and curb climate change by absorbing more carbon than traditional agricultural practices. It also avoids the practice of converting forests into monoculture agriculture, thus maintaining soil and biodiversity.

Subsistence Farming. The majority of the world's agricultural land is neither managed nor produced by industrial agriculture, but by 500 million families worldwide and distributed all over the globe! In fact, these 500 million families produce most of the world's food.[vii] These 500 million people engage in **subsistence farming**—farming to eat, or growing one's own food. It is important that subsistence farming maintains its status as the majority of food production. For one, it is important for food security, so individuals, especially in developing countries, do not rely on exterior sources for food. By growing their own crops, they hold on to valuable knowledge and maintain independence. Second, subsistence farming helps protect the natural environment by avoiding the adverse environmental effects that come with industrial faming.

EcoPrint/Shutterstock.com

Subsistence Farming in South Africa

Case Study

The Real Price of Palm Oil: Spotlight on Endangered Orangutans

Existing for roughly 20 million years since the Miocene era, Tropical Rainforests are among the oldest and biodiverse ecosystems on the planet. They also happen to be home to our arboreal primate relatives, the orangutans. We split with this great ape from our evolutionary tree 10–12 million years ago and share roughly 97% of our DNA with them. The whites surrounding their irises give them a familiar, almost human gaze, reminding us of our recently shared evolutionary past.

Kjersti Joergensen/Shutterstock.com

Their home range includes the Tropical Rainforests of Borneo, Sumatra, and Southeast China. They are the only great ape to spend most of their time—about 95% of it—in the canopies of trees. Their birth intervals are roughly 7–9.3 years, which are the longest of any mammal. This long gap between births means that maternal care is substantial and mothers place a lot of time and

(*Continued*)

energy into child rearing. In fact, orangutan babies nurse for up to 7 years and are carried by their mothers until they are 5 years old. The bond between mother and baby is strong and impenetrable.

Rich Carey/Shutterstock.com
Deforestation in Borneo

Sadly though, our red ape relatives are in big trouble. 90% of their rainforest habitat range—from Sumatra to Borneo to lower China—has been disturbed or degraded, most of it in the last 20 years. Because of mass destruction to their native habitats, orangutans are now in the IUCN-listed Endangered Species category and in dangerous risk of going extinct in the near future.

Human activity is responsible for most of this disturbance and destruction. Legal and illegal logging, mining, and unsustainable palm oil plantations all cut down and destroy the rainforest and therefore the home ranges of orangutans. Our unsustainable use of palm oil, however, remains the biggest contributor to rainforest destruction in this region.

Palm oil is found in many products that we use every day: cookies, chips, laundry detergent, ice cream, candles, cosmetics, toothpaste, etc. The demand for palm oil has caused plantations to increase in size—sometimes illegally. Palm oil needs specific soil that is found in rainforests, so monoculture plantations are created in areas that were once rainforest. Making room for palm oil means slash and burn agriculture—this involves cutting down the forest and burning the debris and litter that remains. A 2018 Purdue University study concluded that palm oil labeled as 'sustainable' is also harmful to the environment and responsible for deforestation.

This has driven many orangutans out from their canopy home and toward great danger. Often times they end up burned, shot, starving, motherless or dehydrated and too often near death.

Currently, there is moratorium on rainforest clearing for palm oil plantations in Indonesia and strict laws in place regulating their size. But such laws are difficult to enforce and corporations and plantation owners often take the risk since great wealth comes from palm oil. But clearly, something must be done.

In response to the dire situation of orangutans, Dr. Birute Mary Galdikas founded the Orangutan Foundation International (OFI). Dr. Birute Mary Galdikas works in the Rainforests of Borneo. She was chosen by Dr. Leakey to study orangutans and has been doing so since 1971. Along with Jane Goodall and Dianne Fosse, she is known as a "trimate," one of the three notable women doing research on great apes.

OFI works with Indonesian Government to rescue, rehabilitate and reintroduce orangutans into the wild. They are often quarantined, given medical care, and eventually returned to the forests where they are reintroduced once they become healthy and strong enough to live independently. OFI also helps patrol the intact rainforest to prevent further expansion of rainforests or mining or illegal logging.

But they still need support and action from the rest of the world to save our beloved orangutans. What we can all do to help: Assist with tree planting programs in southeast Asia (and everywhere for that matter); Come see orangutans at Camp Leakey in Borneo; Promote ecotourism in Borneo and stay in touch with OFI (www.orangutan.org); and most importantly, significantly reduce palm oil use or cut it out altogether.

Source: Sustainable palm oil may not be so sustainable.

Review Questions

1. What is IPM and why is it used? Describe its different components. What are some pros and cons of IPM?

2. Given the types of pesticides we discussed, which type, if any, would be used in organic farming?

3. Describe how genetic engineering works.

4. List 5 pros and 5 cons of genetic engineering

Discussion Questions for Class

A. What are some everyday products of biotechnology?

B. How might the use of transgenic crops contribute to environmental problems?

C. Discuss the pros and cons of growing and buying organic produce.

D. Discuss the pros and cons of GMOs.

E. Discuss cross-pollination and how GMO crops affect neighboring "organic" crops and organic farmers. Who is responsible in this situation?

F. This is a suggested link: http://www.ams.usda.gov/AMSv1.0/getfile?dDocName=STEL DEV3004446. Ultimately, it gives the guidelines for labeling a food organic.

Your Personal Sustainability Journal

Do your own due diligence. Educate yourself! Understand what GMOs are and costs and benefits associated with them. Be able to have an articulate conversation regarding GMOs and make informed personal decisions.

Become a climate-friendly and environmentally friendly gardener. Use practices that are low emissions—use rakes, compost, use a hand-push mower. Recycle yard waste and food waste by composting. Plant cover crops to help maintain nutrient content in the soil. Some of these cover crops are alfalfa, pea plants, and beans. Plant shrubs and trees (they take in and store the most carbon dioxide.)

Support local food systems; Shop at farmer's market whenever possible. Eat seasonally and locally. When shopping at the supermarket, buy local and organic produce. Support Farm-to-school agriculture. This helps spread awareness about eating a balanced and fiber and nutrient rich diets. It also helps local organic farmers. Patronize roadside stands and u-pick operations (orchards where you pick your own apples, blueberries and strawberries.)

If you have a patch of land (or even a window box), plant native vegetation such as alfalfa and clover, that are a good source of food for honeybees.

Try not to buy food or other goods that contain palm oil (real the label).

Buy foods in bulk and without individual containers.

Use "green" manure or compost in your garden.

Look into buying silicon baking mats. They will help you cut down on use of parchment paper and aluminum foil.

Add your own suggestions here.

Endnotes

i. www.cbd.int
ii. Source: IPCC 2007
iii. For further reading, please see VanEngelsdorp. D, Tarpy. D, Baylis. K, Spivak. M, et al. 2012. The Bee Informed Partnership: Using beekeepers' real-world experience to solve beekeepers' real-world problems. *Amer Entomologist* 58: 116–118.
iv. For a "Ted Talk" regarding this, please see http://www.ted.com/talks/marla_spivak_why_bees_are_disappearing.html
v. From www.fao.org/soils-2015
vi. http://www.organic-world.net/news-organic-world.html?&tx_ttnews%5Btt_news%5D=249&cHash=c707669082121788438872f93e472d31
vii. www.FAO.org
viii. Roberto Cazzolla Gatti, Jingjing Liang, Alena Velichevskaya, Mo Zhoua, Science Direct, Volume 652, 2/20/19, pp. 48-51

Chapter 12

Climate Change

DID YOU KNOW?

Dr. Garen Baghdasarian, a marine biologist at Santa Monica College, conducts NSF-funded research on the coral reefs in Southern Taiwan. He and his research team have been collecting samples of *Symbiodinium* algae (flagellate algae), which form symbiotic relationships with corals, giving corals their brilliant spectrum of colors.

The team discovered that the corals can potentially form symbiotic relationships with hundreds of different algae, many more than had previously been presumed. The algae display varied physical responses to environmental conditions and varying degrees of resistance to coral bleaching (coral bleaching is a sign of ecological stress). The team examined environmental stress variables that included varying water temperature and CO_2 concentration (CO_2 forms carbonic acid when it reacts with water, causing acidification—a drop in pH, refer to Chapter 2 for review).

Dr. Baghdasarian and his research team have concluded from their research that *an increase in temperature* is the leading cause of coral bleaching. Change in pH *may* have long-term physiological effects on the coral but *no discernible* short-term effects.

What Is Global Warming?

Scientists have become quite effective at determining that which is a normal fluctuation in the earth's climate pattern, and that which is significant changes in climate caused by an external driver (such as greenhouse gases). Please consider this contextual example: If you are exposed to a pathogenic strain of bacteria, your body and your immune system has the ability to fight it off quite effectively. Similarly, the earth has built-in feedback mechanisms that can help bring it back to balance if its temperatures fluctuate. For example, various organisms have the ability to cloud seed, which creates protective cloud cover in the event that the organism's ecosystem gets too hot. The earth often responds to excess heat with the **Planck Response**, the ability to release infrared energy out of the upper atmosphere if the earth gets too warm. This helps maintain global temperature stability.

But just as you will likely fall ill after being exposed to overwhelming amounts of pathogenic bacteria, so too will the earth be unable to protect itself from overwhelming amounts of external drivers causing changes to earth's atmosphere and ecosystems. Some of those drivers that overwhelm the earth's feedback mechanisms include a profound increase in greenhouse gases such as methane, carbon dioxide, water vapor, and nitrous oxide into the lower atmosphere.

And as we discussed in earlier chapters, *chemical change* always happens initially, then *physical change* happens in response to the *chemical change*. This applies to our atmosphere as well as aquatic environments. We are seeing significant chemical and physical changes in our atmosphere and oceans. We will be examining the effects of this later in this chapter. In addition, our oceans have become incredibly warm, absorbing excess heat-on top of its chemical changes such as increased levels of carbon dioxide. Since our oceans help regulate and moderate climate, there are profound effects to climate as a result of this. The effects include what is happening to our global climate patterns.

In 1988, the United Nations established The Intergovernmental Panel on Climate Change (**IPCC**), a panel of volunteering scientists, adhering to the strict UN guidelines, who address the escalating concern of earth's increasing global temperatures. The panel consists of 2,500 scientists from different countries performing peer reviewed research on climate change, much of it by examining ice cores.

The IPCC determined with 90% statistical confidence that earth's lower atmosphere is getting warmer, and advanced the prediction that global extremes in weather are increasing worldwide, i.e., higher temperatures in areas accustomed to moderation, and milder conditions in places accustomed to extremes.

Furthermore, the IPCC concluded that CO_2 levels are higher now than ever in earth's last 800,000 years. Consider that, in preindustrial revolution times, CO_2 concentrations were under 300 ppm (parts per million) and are now just under 400 ppm. They also estimated that carbon dioxide will remain in the atmosphere from anywhere between 5 and 200 years.

The IPCC concluded that these changes are the result of the combustion of fossil fuel. After reaching these conclusions, the panel agreed that the time has come for the international community to get involved and take drastic measures to reduce carbon emissions in order to slow earth's increasing temperatures.

The following list is a synthesis of climate change detection results by the IPCC regarding global warming. All results have high statistical probability attributing them to anthropogenic causes:[i]

Surface temperatures have increased in the oceans.
Earth's global temperatures have experienced more warming over the last 50 years.
Surface temperatures have increased on every continent except Antarctica.
Temperature extremes have changed.

The troposphere has increased in volume, most likely due to an increase in greenhouse gases.
The troposphere is getting warmer.
During the latter half of the 20th century, the upper several hundred meters of the ocean have warmed.
Sea levels have risen during the latter half of the 20th century.
In response to enhanced greenhouse gas force, there has been an increase in the intensity of cyclones since the 1970s.
Heavy precipitation and heavy rainfall have increased in the latter half of the 20th century.
Reduction of sea ice in the northern hemisphere.
The widespread retreat of glaciers during the 20th century.

Many other scientists affiliated with research institutions are conducting research on climate change and range shifts of plants and animals corresponding with the increase in global temperatures. Peer-review research has concluded that various species of plant, butterflies, birds, mosquitoes, and marine animals are shifting their range either to increased latitudes, altitudes, or to the equator to find amenable temperature ranges (see Chapter 9). Further research has examined the increased occurrence, threat and destruction of wildfire in the American West as a result to an increase in atmospheric temperatures. (See the Case Study at the end of this Chapter.)

UNFCC

In response to the climate research, reports, findings and conclusions if the IPCC, the United Nations created the United Nations Framework Convention on Climate Change (**UNFCC**). The UNFCC is the nexus for action to fight climate change and its profound impacts on people and the environment. It also establishes frameworks for agreements among countries that have ratified the UNFCC. Participating countries meet yearly to track progress and share technologies that would effectively help reduce greenhouse gases, especially for transportation, energy, agriculture, and waste management. The Kyoto Protocol, Montreal Protocol, and Paris Agreement are all a result of the multilateral efforts of the UNFCC to combat climate change.

Kyoto Protocol

December of 1997, an UN-sponsored convention linked to the United Nations Framework Convention on Climate Change. Participants gathered in response to climactic change. Parties met in Kyoto, Japan, to address escalating concerns of climate change. Its goal was to discuss, devise, and enforce measures for industrialized nations to monitor greenhouse gases and carbon emissions, with a goal to reduce these emissions by 5% (relative to 1990s standards which, due to increased carbon dioxide concentrations, translates to almost 30% reduction by present-day standards—gas inflation!!!).[ii] It placed the most responsibility on developed nations, recognizing that they are the most responsible for Greenhouse Gas Emissions as a result of more than 150 years of industrial activity. The negotiated agreement became known as the **Kyoto Protocol** and was not enforced until February of 2005.

Most industrialized nations, to date, adhere to the Kyoto Protocol and its regulations. A conspicuous exception is the United States (which emits 20% of the world's carbon emissions). Shortly after taking office, George W. Bush decided not to support the Kyoto Protocol, as it gave China and India exemptions to allow them to boost their struggling economies. Bush set up individual standards for the United States which ultimately fell short of the criteria or the standards of the Kyoto Protocol.

In recent news, some politicians have declared that global warming is not our problem (meaning the United States) because if we start a campaign to reduce the effects, it would not unilaterally apply to every industrialized nation and would effectively destroy our economy by rendering the United States less competitive, compared to nations without regulations.

Regardless, the Kyoto Protocol is seen by many as a step in the right direction for a global unity and commitment to Greenhouse Gas reduction. And it also provides a framework for future international agreements on climate change.

Given all this talk of global warming, let us now examine what global warming means, what causes it, and how to mitigate its impacts.

Paris Agreement

In November 2015, members of the United Nations Framework Convention on Climate Change (**UNFCC**) met in Paris to discuss how to fight climate change and how to take aggressive measures to ensure an international low carbon future. Participating countries wrote agreements to keep temperature rise below 2 degrees Celsius; pre-industrial levels. The agreement also proposed the creation of more carbon sinks such as forests. And finally, they formed a consensus regarding how to help LDCs most hit by climate change, and how to mitigate their adverse effects through the support of MDCs.

Who Is Responsible for Climate Change?

According to the United Nations Department of Economic and Social Affairs, MDCs released 31,372,021 metric tons per capita of carbon dioxide in 2010, while LDCs released 13,517,325. In 2010, the countries that released the most global Carbon Dioxide were (in decreasing order): China at 23%, the United States at 19%, and finally India, Russia, and Japan. The United States contains only 5% of the World's Population, but according to this report, is responsible for nearly 1/5 of global carbon emissions! This imbalance suggests that each person residing in the United States has a disproportionately large carbon footprint relative to the rest of the world's population. This further suggests that our use of fossil fuels, consumerism and lifestyle here in the United States is not isolated, but rather greatly impacts the rest of the World. Island nations such as the Maldives are most threatened by climate change as they experience rising water levels. Countries with high biodiversity such as Indonesia and Brazil are also facing looming threats of biodiversity loss due to climate change.

What obligation does each of us, here in the United States, have to the rest of the world? What obligation do we have—especially to countries most vulnerable to climate change? How do we have to mitigate the effects of global warming and make reparations?[iii]

Earth's Atmosphere

The **atmosphere** is a compilation of gases that surround the earth. The atmosphere simultaneously filters out UV light and keeps earth's surface relatively warm. We call this the greenhouse effect. The glass panes of greenhouses perform a similar function on a much smaller scale, giving the effect its name.

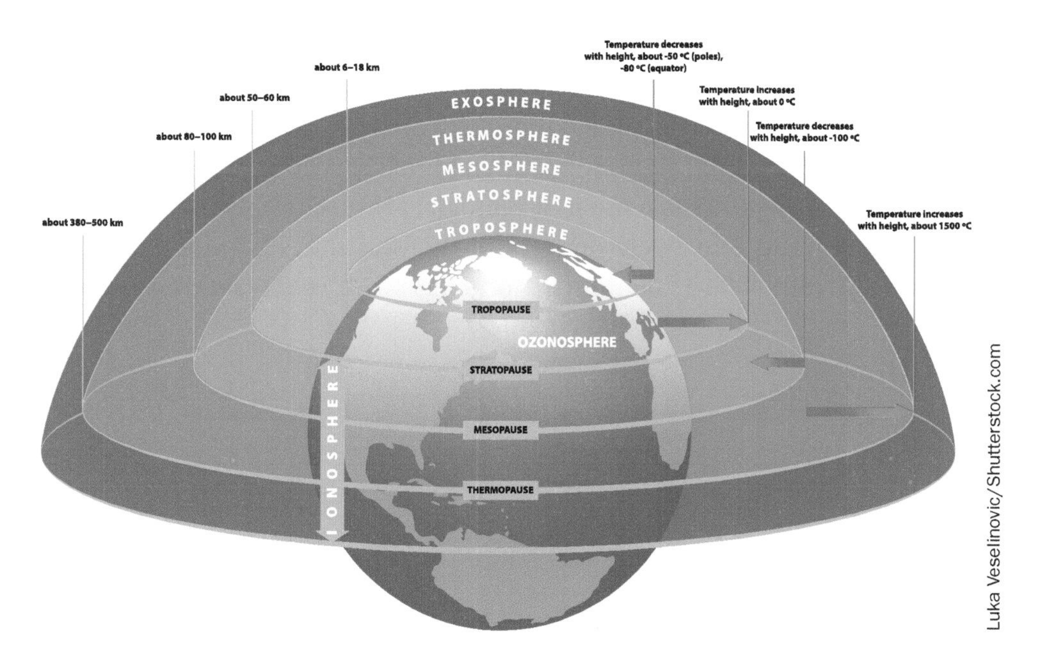

Luka Veselinovic/Shutterstock.com

The atmosphere is further classified by four distinct vertical layers, all varying in composition, temperature, and distance from earth. The four layers are, in increasing distance from the earth's surface, the troposphere, stratosphere, mesosphere, and thermosphere.

The **troposphere** is the atmospheric layer closest to earth and extends from earth's surface to roughly 10 miles outward. The troposphere is made up of gases such as oxygen, carbon dioxide, and nitrogen. The troposphere layer of the atmosphere is expanding. This is more likely caused by an increase in heat-trapping gases.

The atmosphere's second layer, the **stratosphere**, extends from 10 to 30 miles outward. We occasionally enter the stratosphere when we engage in air travel. The stratosphere is significant because it contains high concentrations of ozone (O_3), which prevents UV light from entering earth's lower atmosphere.

Further out we find the **mesosphere**, which extends some 30–50 miles up and sustains temperatures as low as –90°C!

The **thermosphere** is the largest part of the earth's atmosphere extending from 50 to 300 miles outward from earth's surface. The temperature in this layer can reach up to 100°C because it absorbs much of the sun's short-wave radiation (hence the name thermosphere). As a result, many of the molecules in this layer are ionized—which we may witness as the radiant colors of auroras (the northern and southern lights.)

Light is electromagnetic energy which moves through space in **waves** (and as particles called photons). We measure the distance between the crests of these waves as **wavelengths**. Wavelengths can be relatively short, such as gamma waves, X-rays, and UV. Short wavelengths carry a great deal of energy and have been known to alter DNA and cause mutations in humans as well as other living organisms. This also explains why you wear a lead sheet for protection when getting X-rays done. UVA and UVB have been linked to skin cancer. Relatively long wavelengths include long-range light such as microwaves or radio waves.

Long-range light (long wavelengths) emitted from the sun is called **infrared** and carries less energy than shorter-range light. Infrared light helps various organisms thermoregulate by moving or increasing the movement of biological molecules. However, long-range light is not strong enough to effectively drive photosynthesis. PAR (photosynthetically active radiation) ranges from 400 to 700 nm (nanometers). Infrared wavelengths extend beyond 750 nm.

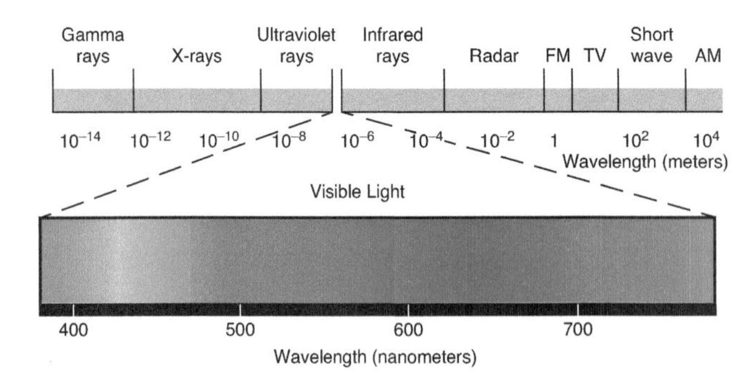

Table of Contents Visual Stimulus

Of the solar energy that reaches earth, roughly 30% is reflected back up into the upper atmosphere and 70% is absorbed into the earth's lower atmosphere, scattered as infrared radiation. Some of this reflects back to earth's surface while some of it escapes into space. Greenhouse gases (such as water vapor, methane, carbon dioxide, ozone, chlorofluorocarbons [CFCs], and nitrous oxide) trap this infrared radiation as heat, reradiate it back to earth, and contribute to the overall warming of earth's surface. The above mentioned gases are **greenhouse gases** and this is known as the **greenhouse effect**.

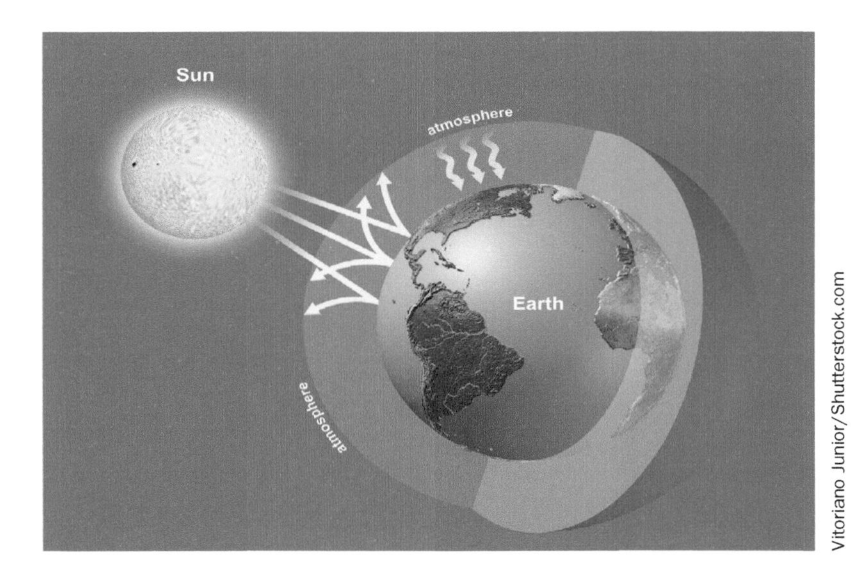

Vitoriano Junior/Shutterstock.com

Greenhouse Gases

Many of the greenhouse gases exist naturally in the troposphere, but human activity has dramatically increased their concentrations and the overall greenhouse effect. Let us examine each greenhouse gas individually.

Watering of lawns, irrigation of crops, and increased runoff have amplified the amount of *water vapor* in the atmosphere (water vapor is the most abundant greenhouse gas).

Methane gas is produced by microbes that break down organic matter. It is found in the ruminant stomach of cows to help cows digest cellulose (grass), and this counts for the majority of methane emissions. It is also found in coal beds and landfills. Our overconsumption of meat, increased use of fossil fuels, and overproduction of waste have increased the amount of methane gas in the atmosphere. Methane accounts for 14% of global greenhouse gas emission.[iv] It is 20 times more potent than carbon dioxide in a 100 year span, and 84 times more potent than carbon dioxide in a 20-year span. Methane does not have the persistence that carbon dioxide has, and so does not stay in the atmosphere as long. So, reducing methane emissions would have a quick and profound effect on climate change that methane is found naturally in oceans, swamps, wetlands, and marshes.

Burning of fossil fuels has increased the amount of *carbon dioxide* in the atmosphere. Furthermore, by examining ice cores, a UN panel of scientists on climate change has been able to get a rough estimate of the concentration of carbon dioxide in the atmosphere through time and have concluded that its concentrations have increased dramatically as a result of combustion of fossil fuel. Carbon dioxide released by combustion of fossil fuel accounts for 57% of global greenhouse gases, while carbon dioxide released by means of forest fires and as a by-product of decomposition accounts for 17% of global greenhouse gas.[v]

Trends in Global Emissions

Global Carbon Dioxide (CO_2) Emissions from Fossil Fuels 1900–2008

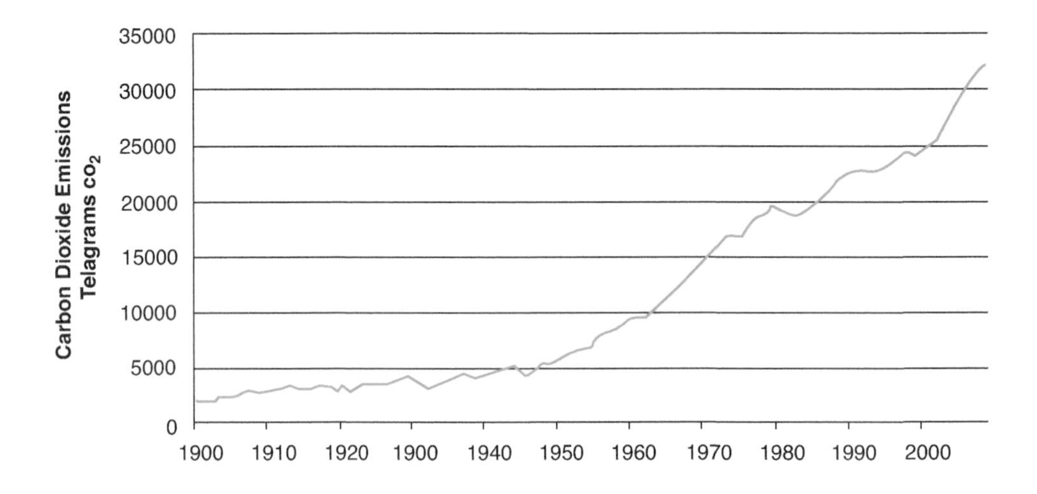

Carole Castelli/Shutterstock.com

CFCs are compounds used in aerosols and refrigerators. They are nonreactive (inert) when found in the lower atmosphere, but eventually, CFCs work their way up to the stratosphere where they are exposed to UV light. When exposed to UV light, they break down and release chlorine, which is known to destroy ozone.

HFCs (hydrofluorocarbons) are Greenhouse gases usually used to replace CFCs and other refrigerants, aerosols, fire retardants and solvents—compounds that are now being phased out as a result of the Montreal protocol because they too harm the ozone layer.[vi]

Any molecule containing Fluorine does not have a natural origin on earth, so were most likely made in a laboratory setting. Compounds with Fluorine typically have high **persistence;** they stick around the atmosphere for an extended length of time—anywhere from 1 to 270 years. HFCs are released through industrial leaks or poor chemical waste disposal, and only go away when the sun's rays eventually break them down. As of 2014, the EPA is considering banning some of HFC's uses. Fluorinated gases constitute 1% of global greenhouse gas emissions

Nitrous oxide forms as a result of our overuse of fertilizers which contain nitrogen. When the fertilizer breaks down, nitrogen reacts with atmospheric gases to make nitrous oxide. Nitrous oxide can also be a by-product of the combustion of fossil fuels. Nitrous Oxide accounts for 8% of global greenhouse gas emission.[vii]

Ozone, as mentioned in previous chapters, is a pollutant found in urban areas as a result of fossil fuel combustion: SF_6 – Sulfur Hexafluroide, has a high persistence and is produced by the electrical power company.

Pollution and Cloud-Cover Effect Global Temperatures

In order to form, clouds need the presence of tiny particles called aerosols. Aerosols can occur naturally, like dust or sea salt. But aerosols can be human caused such as soot. Cloud droplets condense and grow as large as the aerosols allow. Different types of clouds—and the amount of cloud cover—can have an effect on earth's temperature. High, thin wispy clouds such as cirrus clouds can trap heat, deflect it back to earth and have an overall warming effect on earth's surface temperatures. Thick, low clouds such as nimbus or cumulous clouds on the other hand may deflect the sun's rays back into the upper atmosphere, having an overall cooling effect on earth's surface temperatures.

Cirrus Clouds

Recent research suggests that man-made pollution can also affect earth's temperatures by influencing the presence of cloud cover. Pollution particles released from car exhaust and industrial centers (for example) may cause cloud formation, which in turn can affect climate—and may in some cases include an overall cooling effect on earth's lower atmospheric temperatures. Aerosols and particulates can deflect sunlight, causing overall cooling effect on earth's lower atmospheric temperatures.[viii]

alybaba/Shutterstock.com
Nimbus Clouds

Similarly, natural pollution can have an overall cooling effect as well. Volcanoes release sulfates that react with water vapor to form clouds of sulfuric acid. These droplets absorb sunlight and deflect or scatter it into the atmosphere. Earth's temperature during times of great volcanic activity has been substantially cooler than times of less volcanic activity.

The Ozone Layer

Ozone (O_3) is found in high concentrations in the stratosphere where it acts as a shield and filters out almost 99% of the sun's harmful UV rays and keeps them from reaching earth's surface.

As mentioned earlier, CFC molecules are broken up by UV light in the stratosphere, ultimately breaking down ozone and compromising the ozone shield, allowing harmful UV light into earth's lower atmosphere.

In the past two decades, the ozone has thinned worldwide. While limited exposure to UV is beneficial—even essential—to humans, the amount that the thinned ozone layer allows can compromise human immune systems, increase intensity of sunburns, cause skin cancer, and increase the likelihood of cataracts.

Overexposure to UV light also makes plants more vulnerable to parasites. Algae, for example, are vulnerable to UV, and exposure may affect marine food chains and cause a decline in amphibian populations. Ozone negatively affects photosynthesis, decreasing yield, causing pigmentation in plants and stunts plant growth.

In the mid-1980s, the international community took measures to slow ozone depletion. In 1987, an international panel met in Montreal, Canada, to sign the **Montreal Protocol**, a plan of action to cut down on chlorine-containing chlorofluorocarbons (CFCs) and bromine-containing halons, which has been carried out with considerable success. Today, compounds that deplete the ozone are decreasing. The 2014 levels of these substances were 10% below their peak concentrations in 2000,[ix] but areas around the Arctic are still losing ozone. Scientists are still examining the cause for this.

Effects of Global Warming

The results of global warming are many, they are unpredictable, and they are difficult to quantify. Nevertheless, here is a list of some of the more clear-cut consequences.

Infestations of mosquitoes and other disease vectors have increased, resulting in a spike in vector-borne diseases such as malaria and yellow fever.

Plants may flower at different times than their pollinators, preventing their proliferation. Plants also may become more vulnerable to parasites.

The permafrost has been melting, exposing dark-colored ground. Dark colors absorb more wavelength of sun so global temperatures increase even more, causing an insidious positive feedback loop.

Heat waves may become more prevalent, accompanied by increased and more severe droughts and wildfires.

Photosynthetic rates decrease with extreme temperatures.

We may see a disruption of species home ranges, in which species move to higher altitudes or latitudes to seek out optimal temperatures. This may introduce them to new diseases, competitors, and predators, compromising their viability.

Ice caps have started melting, increasing the sea level and submerging islands and low-altitude land. Polar bears, a threatened species, perch in ice caps in order to hunt.

Dominik Michalek/Shutterstock.com

Melting Permafrost

As a result of the ice caps melting, polar bears are being observed swimming for miles, growing tired and fatigued, and eventually starving or drowning. Or, for an alternative food source, polar bears are often forced to hunt larger and more dangerous animals such as walruses—a risky predation. In some areas, polar bears, which traditionally eat marine animals, are observed shifting to a land-based diet. Current research indicates that polar bears have resorted to eating snow geese, mushrooms, and berries. This observation suggests that they are adapting to changes in climate.[x]

Increases in water temperatures have been observed, causing coral bleaching. Coral bleaching compromises the coral reef ecosystem, and many countries rely on ecotourism from the beauty and varied colors of the reefs. Furthermore, as a result of increased water temperatures, amphibian populations are declining.

We may experience an increase in accidents and deaths associated with flooding.

Countries could experience famine as farmers lose their ability to subsistence farm.

The amount of wildfires may increase.

As a result of extreme environmental conditions, marine and land ecosystems could collapse.

Areas covered with glaciers—like Iceland and possibly Alaska—may get higher as the continental plates lift upward, in response to the glaciers melting.xi

Allergy season is increasing! Pollen production for allergens like ragweed is ext-ended, sometimes 27 days longer than average.[xii]

Increase in electricity demand in response to more extreme weather patterns.

Loss in Sierra snowpack.

The troposphere grows in size as the amount of greenhouse gases accumulates.

An increase in tropical typhoons, hurricanes, cyclones, and more intense high tides.

Coral Bleaching

In other words, global warming causes many uncertainties. It is unknown how species, and which ones, will adapt. As you can see, the fickle and indefinite nature of global warming renders planning strategies to deal with it and implementing these strategies controversial and contentious.

What Do We Do about Global Warming?

One proposed solution is to capture carbon and store it as a carbon reservoir in an alternative area to the atmosphere. One way to achieve this is to capture carbon directly from coal power plants and inject it underground into sedimentary rock.

Another means to capture carbon is to stop deforestation and grow more tree farms and roof gardens in metropolitan areas to capture CO_2. Roof gardens not only capture CO_2 but also absorb most rainwater and therefore decrease runoff. They also act as natural insulation for homes, requiring less energy for heating or cooling, reduce noise pollution, and help regulate the hydrologic cycle.

And yet another means to capture carbon involves, of all things, mushrooms! A study published in Science magazine recently suggests that mycorrhizal fungi can help capture and store carbon in soil deep underground.[xiii] Mycorrhizal fungi, the subject of the study, form a mutualistic symbiotic relationship with plant roots to help plants absorb nutrients more efficiently from the soil (these fungi were instrumental in assisting angiosperms colonize land). Soil acts as a large carbon storehouse. When decomposers in soil break down dead plant matter, they release carbon dioxide into the atmosphere as a by-product. These decomposers, however, need soil nutrients, such as nitrogen, in order to survive. According to research in Nature magazine, mycorrhizal fungi compete with these decomposers for nitrogen. If successful, the fungi absorb nitrogen and release it to the plant (with which it is symbiotic). It prolongs the life of the plant, curbing the decaying process and performing long-term carbon sequestration. The mycorrhizal fungi also act as decomposers themselves, and when they break down dead and decaying plant matter, they release carbon deep into the soil, so it is not released into the atmosphere, effectively reducing the amount of atmospheric carbon dioxide.

Researchers recently discovered a catalyst that can, with high efficiency, convert carbon dioxide (a greenhouse gas) into carbon monoxide—which, in turn, can be used to create useful chemicals.[xiv] It is possible that in the future, these catalysts could be used at coal power plants and refineries through renewable energy sources (such as hydro or wind). Human innovation may continue on this trajectory, and we may possibly discover more ways to covert greenhouse gases into a less destructive and even useful substance.

Alison Hancock/Shutterstock.com
Roof Garden on an Urban Building

Another approach is to cut down our carbon emissions altogether. We can do this by using alternative renewable fuel sources discussed in Chapter 6 that emit less carbon and methane; we could invest in and start using renewable fuels and improve technology so alternative energy sources are more efficient; invest in optimizing and getting the energy grid up-to-date so it is more efficient. Another possible solution is to invest in algaculture; using algae to fertilize oceans, capture CO_2, and provide alternative energy (see Chapter 6) source.

And a final solution is to shift in lifestyle, use alternative transportation like buses, subways, and bicycles, and move toward a plant-based vegan diet—or at least reduce meat and dairy consumption and try to live more sustainably.

Case Study

Western Wildfires Intensify as a Result of Climate Change

In recent years, the Western U.S. has experienced extreme, dangerous, and record-breaking wildfires. Granted, wildfires are a natural and essential part of the ecosystem in the western United States. Periodic cool-burning surface fires have many advantages. They control pest populations, release nutrients from soil, and enable trees to crown sprout and nuts to crack open, increasing overall fertility. Some plants even have flammable oils, an adaptive advantage to ensure that they burn with frequency and harness all of fire's ecological benefits on a consistent basis.

Regardless of the many benefits of fire, since the year 2000, the average area of land burned in wildfires has more than doubled from the annual average from years 1985 to 1999.[xv] New climactic conditions and the increase of human development are changing the course of wildlife fires. The length of fire season is increasing, as is fire intensity to levels that are more dangerous than their historical counterparts. As a result, the danger and risk involved is also increasing.

What are these climactic changes responsible for such a dramatic shift? From 1970 to present day, the average annual temperature in the United States has increased by 1.9 degrees Fahrenheit. This is roughly *twice* the rate of average global warming worldwide.[xvi] Warmer temperatures have many side effects that contribute to an increase in fire. For instance, California has been experiencing extreme drought conditions in the last few years, which has many undesirable side effects. Earlier snow melt and drought conditions dry out vegetation, increasing risk and damage from wildfires. Temperature increase is responsible for beetle and other pest infestation. This infestation causes trees to die, and dead trees act as potential fuel wood, exacerbating the problem.[xvii]

To contribute to the problem even further, the western United States has traditionally employed aggressive fire preventative measures. Aggressive wildfire suppression and preventative measures means that the fire cycle has been suppressed and the benefits of natural fires has been thwarted and ecosystems are out of sync with their historical fire cycles. It also means that a lot of underbrush has built up over the years. The dry underbrush acts as fuel for a potential fire, and this makes the potential for hotter and more intense crown fires much greater.

There is a lot of collateral damage to forest fires such as damage to infrastructure, property, and local economies. Smoke from wildfires causes health problems. It can aggravate asthma, lung disease, and heart disease and cause death and injury. These health risks increase when the smoke travels to urban or densely human-populated areas.

Heavily burned areas lead to soil erosion which puts land at risk of flooding. Furthermore, carbon dioxide gas released from fires causes a rise in heat-trapping carbon emissions. They also increase other air pollution such as particulate matter, ash and soot. Wildfires harm watersheds and infrastructure such as water supply stations, power lines, bridges and roads.

There are some steps to take to prevent wildfire intensity and frequency: We can make an effort to reduce construction, development and human expansion near fire-prone areas; we can build resilience to wildfires in at-risk communities and raise awareness and preparedness; Create and enforce stringent statewide firecodes and building codes and create a fire-proof buffer zone. And finally, we can drastically reduce our fossil fuel use and cut carbon emissions that are the catalyst for climate change!

Digital Media Pro/Shutterstock.com

Review Questions

1. List the layers of the atmosphere from the layer closest to the earth's surface outward. What is the basic gaseous composition of each?

2. How does cloud cover effect climate?

3. What is the significance of the Montreal Protocol? Was it successful? Why or why not?

4. List the common greenhouse gasses, and explain how human activity has led to an increase in their concentrations.

Discussion Questions for Class

A. How have root causes of environmental problems influenced earth's global temperatures?

B. The fact that polar bears have shifted to a land-based diet in some areas suggests that they are more flexible than initially believed, and can perhaps adapt to earth's changing climate better than some scientists previously thought. Is this encouraging research? Why or why not?

C. Consider how earth's temperatures have increased since the industrial revolution, commensurate to carbon emissions. Do you agree with the conclusions of the IPCC?

D. How does pollution affect earth's temperatures? What accounts for this?

E. Do you agree with some politicians' viewpoint that global warming is not necessarily our problem? Is our economy a more imminent national concern?

F. Please go to the website www.ipcc.ch. In this website, one can find a document drafted by the chairman of the IPCC, Dr. Pachauri. According to Dr. Pachauri, "Fossil energy use is responsible for about 85% of the anthropogenic CO_2 emissions produced annually." What are your thoughts on this statement?

G. Do you agree with President Bush's decision to not support the Kyoto Protocol? Why or why not?

H. There are two philosophies regarding global warming: (i) Some feel it is inevitable and we need to mitigate its effects and find strategies to store and capture carbon; and (ii) Some feel that global warming is avoidable and we should take every precautionary measure to prevent increasing global temperatures. How do you feel and what should we do about it?

Your Personal Sustainability Journal

Start looking into vegan options. If you have not already done so, make efforts to limit your meat and dairy intake. Dairy and meat use a lot of water and energy, and release methane gas.

Borrow items that you seldom use such as cars, bikes, printers, ladders, and photocopiers, instead of buying them.

Be creative in avoiding harmful chemical pesticides; plant marigolds (natural pesticides), and use soap and alcohol spray.

Make sure your tires are properly inflated at all times. This increases miles per gallon.

Limit use of aerosols and styrofoam as often as possible. Instead of using the air conditioner, use a fan or more natural means to cool a home.

Avoid eating red meat.

Endnotes

i. www.ipcc.ch
ii. http://www.kyotoprotocol.com
iii. The Carbon Dioxide Information Analysis Center Oak Ridge National Laboratory and IPCC, 2007
iv. Source: IPCC 2007
v. Source: IPCC 2007
vi. Source: IPCC 2007
vii. Source: IPCC 2007
viii. David Topping, Paul Connolly, Gordon McFiggans. Cloud droplet number enhanced by co-condensation of organic vapours. Nature Geoscience, 2013
ix. NASA/Goddard Space Flight Center. "2014 Antarctic ozone hole holds steady."
x. D. T. Iles, S. L. Peterson, L. J. Gormezano, D. N. Koons, R. F. Rockwell. Terrestrial predation by polar bears: not just a wild goose chase. Polar Biology, 2013; 36 (9): 1373 DOI:10.1007/s00300-013-1341-5
xi. University of Arizona. "Iceland rises as its glaciers melt from climate change." ScienceDaily. ScienceDaily, 29 January 2015.
xii. Ziska, L., et al. 2011. Recent warming by latitude associated with increased length of ragweed pollen season in central North America. *Proceedings of the National Academy of Sciences, USA*. doi:10.1073/pnas.1014107108.
xiii. Science 29 March 2013: Vol. 339 no. 6127 pp. 1528–1529, pp.1615–1618
xiv. Qi Lu, Jonathan Rosen, Yang Zhou, Gregory S. Hutchings, Yannick C. Kimmel, Jingguang G. Chen, Feng Jiao. A selective and efficient electrocatalyst for carbon dioxide reduction. Nature Communications, 2014; 5 DOI: 10.1038/ncomms4242
xv. National Interagency Fire Center (NIFC) online statistics found at www.nifc.gov
xvi. Climate Central, 2102b. The heat is on: U.S. temperature trends. www.climatecentral.org/news/the-heat-is-on.
xvii. Van Mantgam, P.J., and N.L. Stephenson. 2007. Apparent climactically induced increase in tree mortality rates in a temperate forest. Ecology Letters 10(10):909–916.

Chapter 13
Conservation Biology

DID YOU KNOW?

The original national parks, Yellowstone, Yosemite, and Sequoia, were originally open lands in need of protection from vandals, poachers, and trespassers. In 1872, a group of civilians had been appointed to protect Yellowstone. Despite this noble gesture, they had little legal jurisdiction or firepower to back them up. Furthermore, Congress saw no need to fund protection for such national treasures.

In 1886, the U.S. Cavalry, still without funding from Congress, stepped in and started patrolling these vast lands—first Yellowstone, and later the other abovementioned areas, until 1918 when the U.S. Park Service was implemented. If you ever stop to take note of National Park Service Ranger uniform, he is wearing a modern-day version of a U.S. cavalry hat!

Today the national parks, an agency of the U.S. Department of the Interior, provide multiple recreational purposes for people as well as overseeing plant life and wildlife management. Encompassing 85 million acres, the national parks contain and protect the majority of the U.S. forests, permit logging, mining, and oil drilling, and provide various recreational opportunities, such as hiking and camping.

The National Parks also have a Conservation Association that is committed to protecting wildlife such as wolves and bears and also to reintroduce species into many new habitats in order to protect their population numbers (gray wolves were successfully introduced into the Pacific Northwest for example).

The role of the National Park today is also to take conservation measures in order to protect many different species' from habitat loss—the number one cause for extinction. They are committed to maintaining as pristine an environment as possible to help preserve all species and their respective ecological roles.

The national parks are not without threats. Such threats include pollution and graffiti, crime, eroded trails, invasive species, and disturbance from off-road vehicles.

Conservation efforts are dedicated to preserving species and biodiversity. Currently the world is experiencing an alarming loss of species, loss of biodiversity, and accelerated rates of extinction. We examined Biodiversity and threats to biodiversity in Chapter 3. Now we will examine various ways to preserve and protect individual species and ecosystems, and possible reasons for species decline.

Yosemite National Park

EastVillage Images/Shutterstock.com

Extinctions and the Decline of Species

It has been estimated that 99% of all species that roamed this planet are now extinct.

It is normal for species to go extinct eventually with time. In fact, most species exist for approximately 10 million years, then succumb to extinction resulting from moderate changes in the environment—this is called **background extinction**.

However, due to rapid climate change, overexploitation, deforestation, and general loss of habitat for many species, some experts predict that we are currently experiencing 45 times the normal extinction rate, and that could increase to almost 100–1,000 times[i], resulting in **mass extinction**.

According to the International Union for the Conservation of Nature (IUCN) nearly 22% species of mammals are threatened or going extinct.[ii] In 2014, the IUCN assessed roughly 75,000 species and found over 20,000 to be threatened and almost 7,000 to be endangered.[iii] And our oceans are not faring any better. It is estimated that 20–25% of our common species living in the oceans are threatened with extinction.[iv] This is the same percentage of land plants and animals in the same predicament.[v]

A **biological extinction** occurs when a species is no longer found anywhere on earth. The dodo bird, eastern cougar, and passenger pigeon have all succumbed to biological extinction and no longer walk this earth. **Ecological extinction** occurs when a species' population size becomes so reduced in size, that members of the species become ineffective at performing their ecological role or having significant community relations with members of other species. The guanaco population of the Patagonia region of South America, for example, has reduced drastically as a result of introduced species. Present day, guanacos no longer serve their ecological role as prey for larger species, disrupting the food web.[vi]

If population does indeed get drastically reduced, the species may succumb to the **Allee effect**, which is in essence a self-thinning rule. The Allee effect occurs when species numbers become so low that it has a negative effect on the population size which, in turn, keeps reducing. Population reduction occurs because either members of a species do not have suitable access to mates, or there is ineffective cooperation in food gathering, etc., and the population ultimately experiences a decline in population growth and thus becomes vulnerable to extinction.

Pichugin Dmitry/Shutterstock.com

Guanacos in Patagonia, Chile

gallimaufry/Shutterstock.com

An Extinct Dodo

A **local extinction** occurs when a species is **extirpated**; goes extinct in a specific area but is found elsewhere in the world. Rhinoceros, for example, have gone extinct in Mozambique, but are found elsewhere in Africa—although threatened and almost driven to the brink of extinction.

A **deterministic extinction** implies that a species will go biologically extinct unless things change, such as preserving habitats and conducting aggressive conservation efforts. Species facing imminent extinction are sometimes referred to as a **vulnerable species**. Typically, **endemic species** are vulnerable to extinction. Causes for deterministic extinction include the following:

- **Habitat destruction:** Habitat destruction is the number one cause for extinction. Cutting down trees for livestock grazing and agriculture, as well as building homes and roads are all unsustainable practices that take habitats away from plants and animals.
- These activities also cause **habitat fragmentation,** a dramatic reduction in the size of an area or habitat of a species. Large mammals such as male mountain lions have a habitat distribution of one male mountain lion per *at least* 50 square miles (it is important to note that this is a conservative estimate).
- **Overkill.** This occurs when a species is overhunted or overharvested and few species remain in the wild. Animals that are large, tasty, and slow in movement and have parts perceived as useful to some humans, are naturally vulnerable to overkill. An example is the African Elephant which only procreates every 3–9 years and is killed for its tusks. Fearing overkill and thus the imminent extinction of elephants if drastic measures were not taken, conservationists lobbied for a worldwide ban on ivory. In 1989, their efforts were actualized.
- Bluefish tuna, a desirable type of fish, has been depleted at alarming rates due to illegal fishing. It is predicted that they may become endangered if the unsustainable fishing continues.

Marcelle Robbins/Shutterstock.com

- **Poaching** is the illegal human killing and commerce of animals for their valuable parts. Rhinoceros and elephants, for example, are hunted for their tusks. Lions and leopards are hunted for their beautiful fur coats. Poaching is also responsible for the decline in chimpanzees and gorillas in Western Africa. They are illegally sold as **bushmeat**. Again, these vulnerable animals usually do not leave a lot of offspring, so they are particularly vulnerable to extinction.

- **Introduced species**: Introduced species, also known as exotic species, are introduced into a given area from other parts of the world. They are usually generalists, thriving in a variety of environments, and easily adapted to their new niches. Take into account they are no longer in their natural habitats, which means they no longer have their natural population controls such as predators or competitors. Without such population controls, they become a possible pest by growing in population and driving native species out of their habitats. Introduced species are responsible for 40% of all extinctions. The Nile perch, for example, was introduced to Lake Victoria in Africa, and has caused extinction of 200 species of cichlid fish.
- **Chains of extinction**: All species are connected through a food web. Therefore, it stands to reason that large predators disappear when their prey go extinct (since they lose their food source). For example, the black-footed ferret became endangered when the population of prairie dogs, their prey, began to decline.

MVP, or *minimum viable population*, is the minimum number of individuals in a population of species in the wild that will provide enough genetic variability to sustain its population in the wild. The MVP for mammals is generally about 500 individuals, but this number varies considerably depending on the species. Recent research has estimated that at least 20,000 members of a specific species must be present in their respective population in order for biological evolution to occur.

When a species is about to go extinct, it is listed as **endangered** by the U.S. Fish and Wildlife Service. Many factors could cause a species to become endangered. Often their habitat is compromised or they are too few in number to play an ecological role or have a healthy amount of genetic variability. Northern spotted owls, humpback whales, black-footed ferrets, and African and Asian elephants are all examples of endangered species, but there are many, many more.

According to the **Endangered Species Act (ESA) of 1973**, it is illegal to harm or kill an endangered species or to remove it from its natural habitat. The ESA gave the U.S. Wildlife Fish and Game authority to protect endangered and threatened species both at home and abroad. For example, no endangered or threatened species, nor their body parts, can be imported into the United States without ample proof that in doing so, the well-being of the species improves. A **threatened species** is a species that is about to be endangered in the foreseeable future, and conservation efforts need to be implemented to preserve their habitats for their survival. Twenty five percent of known mammals are considered threatened.

Currently, 25% of all mammals are threatened, mostly due to habitat loss. If extinction rates continue as they are, it is estimated, according to a 2011 study in the journal *Nature*, that 75% species that exist today could go extinct.

Under the Lacey Act of 1900 (amended in 2008) endangered species may not be imported by any means:

> It is unlawful to import, export, sell, acquire, or purchase fish, wildlife or plants that are taken, possessed, transported, or sold: (1) in violation of U.S. or Indian law, or (2) in

interstate or foreign commerce involving any fish, wildlife, or plants taken, possessed, or sold in violation of State or foreign law.[vii]

CITES

Implemented in 1973, **CITES** (Convention on the International Trade in Endangered Species of Fauna and Flora) is an international agreement between 179 countries (referred to as parties) participating on a volunteer basis. The agreement monitors the international trade of wildlife and plants by issuing permits and licensing for those legally imported and exported. If and when they are transported, CITES oversees the process to ensure no harm or injury occurs. Roughly 30,000 species of plants and 5,600 species of animals are protected by CITES. Although parties participate on volunteer basis, agreements are legally binding.

In terms of *illegally* traded animals or poaching, CITES cooperates with governments to fight against wildlife crime and assists in designing strategies to combat poaching. Animals—sometimes those that are endangered—are readily exploited for their valuable parts, making the wildlife trade a billion dollar industry. Despite rigorous efforts and cooperation among the international community, elephant and rhinoceros poaching is at an all-time high. (According to CITES.org, in the year 2000, seven rhinoceroses were poached and in the year 2013, there were over 830.) In 2013, The UN Security Council declared The Central Republic of South Africa in crisis due to its rampant poaching. Given the amount of money at stake (kilogram per kilogram, as of the year 2013, rhinoceros tusk was worth more than gold), poachers have amassed an arsenal of sophisticated weaponry and have been known to open fire on park rangers who try their best to thwart the slaughter of elephants and rhinoceros. In turn, park rangers in South Africa have been receiving training by former British and other International Special Forces, so they now have infantry-style training to fight against the poachers.

Criticism of CITES is that they are not doing enough to fight poaching and protect vulnerable, threatened, and endangered species. Consider that recent research estimates that a staggering 10,000 elephants were killed from the years 2010–2012.[viii] The African Lion population has decreased 80% from its historical range, and conservationists estimate that there are only 24,000 left living on the African Continent.[ix] Rhinoceros are in the worst shape of all. There are only about 5,000 Black Rhinos left off in the wild on the African Continent and 20,000 White Rhinos.[x] Despite these staggering numbers, poaching continues to climb.

Legal government-issued permits are allotted for Trophy hunting, another controversial practice, for all of the aforementioned species. Although trophy hunting is legal, many question the ethical considerations of hunting sensitive and rare species for sport. And many are concerned that trophy hunting is a smokescreen for poaching, as such activity occurs in countries with little governance and with widespread poverty. Such places are unable to create or enforce conservation laws, since many of its inhabitants live in impoverished conditions and all funding and energy goes toward helping its people. The temptation to surreptitiously "up" the lucrative hunting quotas—especially given little oversight from government agencies—may be great.

Case Study

The Akashinga

A legion of female rangers are leading the charge against poaching in the lower Zambezi Valley of Zimbabwe. They are called the Akashinga, or 'Brave Ones' and put their lives on the line every day to protect African animals from the illegal wildlife trade. Many come from disadvantaged backgrounds such as poverty, human trafficking, or abusive relationships. But they have found their calling as well as their independence in protecting these majestic animals.

The inspiration for the Akashinga came from Damien Mander, founder of IAPF (International Anti Poaching Foundation). He is a former Australian special forces commando - turned anti-poaching ranger – turned animal welfare and environmental advocate. He put the women through a grueling selection process and to his surprise, very few dropped out. He is inspired by their commitment.

The Akashinga make excellent wildlife rangers because, according to Mander, they are not as corruptible as their male counterparts. Females tend to be better skilled at gathering intelligence, saving months of surveillance. Females are also good at deescalating tense situations, minimizing the risks of armed conflict that may place local communities in crosshairs. Avoiding conflict helps instill confidence in local communities, ensuring their cooperation with intelligence gathering. Females also tend to put more money into the local communities and economy. This is essential, as the success of local communities in strongly aligned with conservation success.

Having been met with incredible success and support, Mander plans to expand this conservation model throughout the African Continent. The future of conservation could very well be female rangers.

***(Featured Picture is Sergeant of the Akashinga, Vimbai Kumire)

Preserving Species

Conservation of species is either **in situ** or **ex situ**. In situ conservation is done in nature, in species' natural habitats: in parks, wildlife reserves, and protected areas. Ex situ conservation is not performed in natural habitats, but is typically performed in human controlled areas: zoos, botanical gardens, animal parks, seeds banks, captive breeding, etc.

Often times, wildlife conservationists create **wildlife corridors** that connect two patches of wildlife habitat that have been interrupted by human encroachment, such as roads and developments. This makes the separated wildlife areas contiguous, increasing the overall intact land area. Wildlife corridors have succeeded with panthers, bobcats, cougars, and deer. In fact, a large-scale conservation project underway in Florida creates and conserves a wildlife corridor from the Everglades to Atlanta. This ensures that various species have an intact and available means for migration.

Baloncici/Shutterstock.com

Botanical Gardens

Robert Crum/Shutterstock.com

Highway Crossing Bridge for Animals, Banff National Park, Canada

Efforts to create and conserve wildlife corridors are carried out by multiple agencies, such as the national parks, the state park system, U.S. Fish and Wildlife, and the U.S. Bureau of Land Management.

Parks, zoos, aquariums, and botanical gardens often employ conservation efforts designed to sustain a species that is threatened or close to extinction.

Conservation organizations, such as zoos and aquariums, breed species in captivity. Captive breeding for conservation purposes is intended to promote healthy genetic diversity in a population, especially for populations that are near extinction. Over time, many of these species are released into the wild and their progress is monitored closely. This has been successful for some species such as the Giant Panda, Whooping Cranes, and Woodrats.

Unfortunately, the overall success rate is low—only about 10%. Considering the many unknown factors that are involved in reintroduction, it is almost impossible for wildlife and conservation biologists to determine which species will be viable and will thrive after reintroduction. Captive bred animals appear to be missing essential survival skills evident in their wild-raised counterparts. For some reason, this knowledge is neither learned nor passed down ex situ. For example, captive-bred lemurs, native to Madagascar, were raised in California and reintroduced into the wild. Shockingly, several of them turned up dead, having poisoned themselves by sampling all the plants and eating berries—not realizing that they were poisonous to their systems.[xi]

Thus, it seems logical that animals learn what is medicinal and what is harmful by observing members of their own species. For reasons that are still unclear, this knowledge must be learned in the wild, of their native habitat.

Captive bred animals also have numerous reproductive issues that are still not explained: elephants do not display prolonged interest in sex, the testes of male gorillas have been known to shrivel and ultimately become useless, and it is difficult for many pandas to breed. These reproductive obstacles suggest that the stress of living a life of confinement takes a physical toll on captive-bred animals.

Many wildlife biologists believe that we should expand the field of **ethology**—the scientific study of animal behavior. Research done in the wild through observation of animals and their behavior in their natural environment may fill in necessary gaps and help determine what needs to be adjusted or refined in captive-bred environments.

Another possible solution is *Assisted Migration and Colonization*—that is, facilitating the transplantation of a species. This idea was first proposed by Camille Parmesan, an Assistant Professor of Integrative Biology of the University of Texas at Austin. Dr. Parmesan is considered a leader in Conservation Biology as a result of her extensive research on Edith's checker spot butterfly as well as many other species.[xii]

Many see this as a controversial proposition, since risks of assisted migration may include unintended introduction of new pests and diseases. Another risk is that the species may become invasive. Furthermore, species reintroduced to a new area will have lack of knowledge of the surrounding vegetation relative to habitats in which they co-evolved. Thus, they may inadvertently ingest a poisonous plant as seen with the Madagascar lemurs transplanted from California mentioned above. Several other unintended consequences are possible as well.

A way to help preserve endangered plant species is to create **seed banks** (also known as gene banks). Seed banks are storage facilities where seeds are stored in low temperatures and low humidity (to protect them from drying out). Although seed banks offer seed protection from habitat loss (and loss of genetic variability from plant hybridization in the wild), many seeds do not withstand being dried out for long. Furthermore, since seeds are ex situ, they are not in the wild, and therefore do not undergo the mechanisms that would enable them to adapt to a changing environment. For example, there is no natural selection or biological evolution, which would ultimately increase their fitness in their natural environment. Therefore, seed banks are not a viable long-term option.

gary powell/Shutterstock.com

The Checkerspot Butterfly

The Problem with Zoos and Living in Captivity

Zoos have come under fire in recent years as critics say that although they are effective in educating on the value of biodiversity, they have little conservation value. Many of the animals in zoos do not need conservation to begin with, or they are sterile hybrids that render conservation ineffective. Then there is the ethical question of whether it is humane to confine animals to captivity for the duration of their lives.

T.Dallas/ Shutterstock.com

Animals in zoos live longer, but this does not necessarily mean they enjoy better health in their longevity. Many captive animals display typified behavior such as pacing, weaving, and rocking, and have on occasion been known to inflict self-abuse. Many zoo gorillas are infertile, often lose interest in mating, suffer eating disorders, and die from cardiovascular disease. Captive elephants often lose interest in mating as well. A considerable number of captive rhinos suffer and die from hemolytic anemia. These trends are not observed in the wild.

Animals also suffer innumerable health issues because they are not in their natural environment and cannot self-medicate. Their new environment is devoid of the soil, water, microorganisms, food, and climate they are adapted to in the wild, while wound-tending and grooming, observed in the wild and important for survival, become lost behaviors. In captivity, they are exposed to new pathogens and diseases to which they have not yet built up immunity. It is difficult for captive animals to get adequate amounts of exercise or even the right vitamins and minerals (calcium, phosphorous, and vitamin D). Even getting the right amount of sunlight and shade is imperative to wild animal health. Many animals sunbathe in the wild to synthesize vitamin D and to self-medicate (even some bacteria have been known to use sunlight for correcting damaged DNA). Zoos do not necessarily provide ample sunlight and without it, many animals suffer from deficiencies and can have bone abnormalities. For all the above concerns, the trend is to make zoo environments as similar to the species' natural environment as possible, but this has obvious limitations.

Finally, the consumer market seems to be shifting as people are less inclined to be entertained by viewing animals in captivity and prefer to observe them in their natural habitat. This is reflected in circuses deciding to phase-out using animals as entertainment and in the growing trend in eco-tourism.

What Factors Influence the Success of Conservation?

Why are some particular conservation efforts successful while others are not? According to a 2014 study in *Nature*, most marine protected areas do not achieve their conservation goals, while many conservation efforts, such as the Florida Wildlife Corridor initiatives, have been successful. What accounts for this difference?[xiii]

Economic growth may have an impact, albeit complicated, on conservation. Economic growth often means a relative abundance of disposable income, translating to more philanthropic endeavors. This in turn translates to ample donations to nonprofit organizations dedicated to conservation agencies, such as the Nature Conservancy, The Sierra Club, and the NRDC. This is significant since, in the United States alone, nonprofit agencies donate billions of dollars to conservation and preservation of species. In this regard, economic growth may *positively* impact conservation.

On the other hand, economic growth may also lead to unsustainable development of houses, roads, and infrastructure, as well as deforestation as consumer demand increases. This ultimately encroaches on species' habitats, undermining conservation efforts. In this regard, economic growth may *negatively* impact conservation.

Science is a major component of conservation and leads to invaluable information about species and the environment. But there are other factors that contribute to conservation success. Some such factors include planning, education, public relations, and political action. Blogs, Letters to the editor, opinion pieces in newspapers, education, meetings with government officials, activism, petitions, and engaging the general public in all areas of the conservation movement are influential, and sometimes imperative, to the success of a conservation movement. In other words, grassroots efforts are a major contributing factor to the success of environmental conservation.

It is also often more impactful when multiple agencies make jointed efforts toward a particular conservation effort. Agencies such as state fish and wildlife agencies, coastal conservancies, departments of forestry, and state and federal park services are more effective when they work together for the same overarching goal.

Countries with corrupt or little governance may find it difficult to successfully conserve land and preserve its species. Developing countries (as we saw in the case study of Ecuador) may be willing to take money from oil companies and allow drilling for an economic boost to the economy, instead of committing to the long-term efforts to build an economy on eco-tourism or sustainable agriculture.

Corrupt government officials may look the other way when it comes to unsustainable hunting for sport. Or well-intentioned governments may lack the resources needed to successfully combat poaching. This is a contentious issue in Africa where much of its wildlife (elephants, rhinos, lions) is poached, hunted for sport, or raised on farms for recreational hunting. Governments may be unable to curb poaching, or may concede to wealthy hunters, who are willing to pay several hundred thousand dollars to hunt such rare and endangered species.

Namibia for instance has auctioned off the lives of endangered Black Rhinoceros (of only a few thousand left in the wild) to recreational hunters who pay to shoot and kill them in the wild. A better alternative is to keep the animals alive and build an economy through walking safaris and eco-tourism, which has been done with success in Botswana. Eco-tourism is estimated to bring in 15 times more revenue from livestock than does trophy hunting.[xiv,xv] In other words, animals provide more economic value when alive than when hunted (not to mention their innate, intrinsic, and ecological value).

Some of the most effective conservation efforts are community-based. The underpinnings of this philosophy are that effective wildlife conservation and the fight against poaching will not be complete without the consent and the cooperation of the local community. In some conservation areas, many locals themselves are suffering from food insecurity and are living in abject poverty. Wildlife and environmental concerns therefore take a backseat to the welfare of the people (if you recall, poverty is a cause of environmental problems). On the other hand, if conservation efforts assist in breaking the cycle of poverty of the local people though microbusiness, micro-lending and education; then possible extended efforts can be directed toward the wildlife and other conservation concerns. This also allows the opportunity for locals to establish a long-term economy built on eco-tourism. A good example of this Conservation model is LEWA Conservancy, a UNESCO organization that promotes wildlife conservation in Kenya through education, microlending, and community outreach. LEWA is considered by many as one of the most successful conservation models.

An Ecosystem Disturbance and Succession

Ecosystems are always changing—they may be thriving with biodiversity one year but enduring a major disturbance that kills many of its species in the next. A **disturbance** is any event that compromises or causes significant change to an intact ecosystem.

A disturbance may be as mild as a small fire or livestock overgrazing, or as extreme as a volcano. The more biodiversity an ecosystem has, the more likely it is able to resist disturbance and/or to subsequently recover from it (unless of course the disturbance is as severe as the aforementioned volcano or an ice age). A **Primary Succession** occurs in an area where there has previously been no wildlife, no soil—nothing for regrowth. It usually consists of bare rock and low quality dirt. A **secondary succession** typically occurs where the previous ecosystem was compromised or profoundly disturbed in some way. Deforestation, a forest fire, flood, and over-grazing are some examples. How disturbed the ecosystem is will determine what type of growth or plant community flourishes first.

Succession is the sequence of events that occur after an ecosystem disturbance. Succession will continue until a **climax community** is reached. A climax community is a stable ecosystem that is rich in flowering plants, trees, biomass, and biodiversity, in general.

After an initial disturbance, the first organisms to grow are aptly called **pioneer species.** Pioneer species facilitate the growth of other plant species by enriching the soil. Lichens, moss, and lupines (pea plants) are pioneer species that facilitate the growth of other plant species. (Lupines make the soil rich with nitrogen.) Lichens are especially likely to appear during primary succession because they secrete acids which help break up rock and mineral, which eventually become nutrients to sustain life. Eventually, lichens die and they help create and enrich the soil even more.

The cycle of change for an ecosystem's succession is broken into blocks of time. They include the following:

A. **Pioneer growth** (or early growth) spans the first 6–10 years. During this time, lichens, moss, nurse plants proliferate, annual weeds that produce seeds then die, and finally shrubs start growing and facilitate growth for other plant species. Typically during this stage, we find plants that are resistant to the harmful effects of an abundance of sunlight. Eventually, the pioneer growth successfully alters the soil and a microclimate emerges, facilitating growth of new species. The new species eventually move in, many with broader leaves that create lots of shade, and shade-tolerant vegetation.
B. From ~7–15 years, the ecosystem experiences **vigorous flowering** of small plants and shrubs. The species that grow during this stage are drastically different from pioneer species, as the soil and amount of shade cover has changed.
C. From 14–25 years, the ecosystem sustains more mature species with lots of biomass, such as trees and large flowering plants. At this stage, the ecosystem has reached its **climax community**.
D. Finally, the ecosystem enters a **degenerate stage** when much of the biomass starts to die and decompose. This stage produces a great deal of dead, organic matter. Lichens and bryophytes (moss) are common, replacing the more mature plant life.

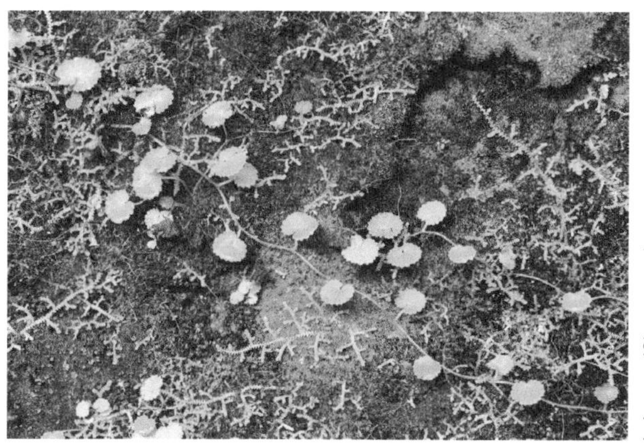
kukuruxa/Shutterstock.com

Pioneer Growth

A **restoration** is the systematic, scientific approach carried out by humans to bring back a disturbed ecosystem to its original state. Many wetlands, coral reefs, and forests have all been disturbed and subsequently restored to their almost-original state by man. The restoration of an ecosystem does not only restore biodiversity by re-introducing native species, but it improves overall soil quality as well.

Rehabilitation is taking a damaged ecosystem and returning it to a state where it is ecologically functional. An example of rehabilitation is the cleaning up of pollution and oil spills so that aquatic habitats are functional. Bioremediation—the use of microorganism metabolism to remove pollutants—is often used to assist in rehabilitation. Although a functional ecosystem can be restored, it is never the same level of biodiversity of an undisturbed ecosystem.

szefei/Shutterstock.com

A Climax Community

Case Study

Costa Rica: A Paragon of Conservation and Sustainability.

Costa Rica, a small country in Central America—home to lush tropical rainforests and vibrant coral reefs—contains 3% of the world's biodiversity. A common practice in many developing countries fortunate to be rich in such natural resources is to engage government subsidies encouraging the sale of private property in order to create open space for plantation crops (like palm oil)

and livestock. Costa Rica's government, on the other hand, firmly believes that Costa Rica's rich biodiversity is a currency and would ultimately be a long-term economic investment. They designed a long-term strategy that would ultimately stabilize their economy and improve overall standard of living. This strategy included preserving biodiversity, not destroying it. The government disposed of their military and started paying private citizens to conserve their land. About 25% of their land became a designated conservation area, thus enabling their economy to be driven by ecotourism.

To preserve the environment and biodiversity—now their biggest currency—Costa Rica proactively promotes sustainability and stewardship. It uses hydroelectricity as major energy source and their other energy sources are renewable as well. Their hotels are even rated on sustainability!! It is not unheard of for some hotels to use recycled gray water for their toilets. This practice has proven itself a remarkable success. Costa Rica's economy and standard of living is much higher than other developing nations.

Eco-tourism, however, is a bit conundrum for Costa Rica. On one hand, the country relies heavily on eco-tourism for its well-being. On the other hand, tourism creates the biggest ecological footprint (with air travel, pollution, etc.). This may explain why the country emphasizes sustainable practice for everybody, including its tourists.

Review Questions

1. Distinguish between background extinction, mass extinction, biological extinction, ecological extinction, local extinction, and deterministic extinction.

2. What are 6 major Causes for deterministic extinction?

3. What is the significance of CITES and the Lacey Act?

4. What health issue do captive animals sustain?

5. Describe the major steps of Succession. What is the difference between primary and secondary succession? Please provide example.

Discussion Questions for Class

A. What are root causes of environmental problems? How have they influenced our need for conservation? How have they increased the amount of ecosystem disturbance and influenced the rate of succession?

B. How has the recent oil spill in the Gulf of Mexico affected aquatic life and marine life in the surrounding area? How involved should the rehabilitation efforts be? Who is responsible for rehabilitation?

C. What efforts can we make in our daily lives to prevent deterministic extinction?

D. Do you think reintroduction efforts of zoos are worthwhile at a 10% success rate? Are zoos ethical?

E. Why are disturbed ecosystems vulnerable to invasive species? Why is an ecosystem with high biodiversity able to resist disturbances?

F. How can CITES be more effective and gain more global cooperation?

G. Why have Costa Rica's conservation efforts been so successful?

Your Personal Sustainability Journal

Plant seeds of native plants or trees.

Plant native plant species in your garden or join a local state or national park for revegetation projects.

Donate time or money to conservation facilities, botanical garden, or wildlife refuges.

To help animals that are overhunted on the African Continent, Check out these websites: http://www.savetherhinotrust.org; http://www.elephantvoices.org; http://www.ewasolions.org; http://lionguardians.org.

Add your own suggestions here.

Endnotes

i. www.unep.org
ii. www.iucnredlist.org
iii. www.iucnredlist.org
iv. http://www.iucnredlist.org/about/summary-statistics#Tables_1_2
v. Thomas J. Webb, Beth L. Mindel. Global Patterns of Extinction Risk in Marine and Non-marine Systems. *Current Biology*, 2015; DOI: 10.1016/j.cub.2014.12.023
vi. Novaro, Funcs, and Walker; Biological Conservation 2000
vii. www.fws.org
viii. G. Wittemyer, J. M. Northrup, J. Blanc, I. Douglas-Hamilton, P. Omondi, K. P. Burnham. Illegal killing for ivory drives global decline in African elephants. *Proceedings of the National Academy of Sciences*, 2014; DOI:10.1073/pnas.1403984111
ix. Bauer, H., Nowell, K. & Packer, C. 2012. *Panthera leo*. The IUCN Red List of Threatened Species. Version 2014.3
x. 31 December by the IUCN for African rhino species
xi. M.L. Drew and M.E. Fowler, "Poisoning the Black and White Ruffed Lemurs by Hairy Nightshade", Journal of Wildlife Diseases, 19(1) (1983) 494–496
xii. www.utexas.edu
xiii. *Nature* 506, 167–168 (13 February 2014) doi:10.1038/nature1305
xiv. Sims-Castley, Rebecca, Kerley, Graham I H, and Geach, Beverley, "A Questionnaire -Based Assessment Of The Socio-Economic Significance Of Ecotourism-Based Private Game Reserves In The Eastern Cape", Terrestrial Ecology Research Unit, November 2004
xv. "Big Game Hunting in West Africa. What is its Contribution to Conservation?" IUCN, 2009

Chapter 14

Human Health

DID YOU KNOW?

As we experience a surge in superbugs—bacteria that have become resistant to the common antibiotics we use in order to supplement our immune response against them—some scientists believe that quorum sensing is our next step in controlling bacterial pathogenicity.

So what exactly is quorum sensing and how does it work?

First, let us introduce the fact that single-celled bacteria can be successful when socially congregating and coordinating action while organized in colonies or living in groups. They communicate through pheromone-like molecules. Pheromones are chemicals that act as long-distance signalers toward other organisms (in this case, toward other bacteria) and help to coordinate population behavior once a certain threshold or population count is reached.[i]

When colonies reach a certain size, bacteria collectively cooperate with each other by turning on genes for pathogenicity or any other mediated behavior that is preferably selected. It is important they reach a certain threshold in size; otherwise the genetic expression for pathogenicity would be negligible and wasteful energy-wise.

Research is showing that quorum-sensing disruption occurs when a synthetic inducer (similar in shape to a natural inducer) is introduced into the colony. The synthetic inducer competes effectively for receptor sites, in effect blocking pheromones (inducers) from the target receptors. This thwarts bacterial communication. Bacteria never become aware of the size of their colony and do not turn on genes for pathogenicity.

Current research has proven that quorum sensing has been used to control MRSA infections. This is optimistic news, since MRSA is a type of *Staphyloccocus aureus* (bacteria) that has become a superbug-resistant to the traditional antibiotics used to control bacterial infections.[ii]

Pathogens

Pathogens are disease-causing agents. The most common pathogens are bacteria, viruses, fungi, or protozoa. When our immune system is compromised or we get a heavy dose of the pathogen, we may fall victim to pathogens and experience multiple symptoms of illness.

An **infectious** or **communicable** disease is transmittable between hosts and passes from one subject to another via various means of contact. Sometimes pathogens such as the influenza virus can infect people who simply breathe the same air or touch the same surface—such as a doorknob.

A **nontransmittable** disease cannot spread through host contact. Rather, it is a genetic disorder that one is born with, or acquires later in life through chance, environmental factors, or lifestyle. Examples include heart disease, diabetes, bipolar disorder, and cancer. A pathogen is typically not the cause for nontransmittable diseases.

In the following section, we will be examining some common pathogens. Notice that some of these are biological contaminants that we discussed in Chapter 8. The reason why they are considered contaminants is that they are potentially harmful to human health.

Bacteria

Bacteria are single-celled **prokaryotic** microbes (microscopic cells that cannot be observed with the naked eye). Bacteria are alive, and so, like all living things, are comprised of organic molecules (proteins, lipids, carbohydrates, and DNA). Bacteria do not have nuclei and are smaller and less complex than eukaryotic cells that make up complex multicellular organisms (like animals and trees).

Bacteria are often observed living in colonies, which suggests social cooperation is common between individuals to maximize survival. Such cooperation is necessary for quorum sensing, which we discussed above.

Bacteria are sometimes ensheathed in a cellular capsule that assists them in lodging on mucous membranes of its animal host and they may also use flagella for mobility.

Salmonella and tuberculosis are examples of pathogenic bacteria. Strep throat is attributed to bacterial infections (*the streptococcus bacteria*), and some skin and mucous tissue infections may be resultant of opportunistic staph populations (*the staphylococcus bacteria*).

In recent news, we are experiencing an emergence of super-bugs (MRSA, VRSA, and TB, among them) due to the increasing number of bacteria that are becoming resistant to traditional antibiotics. This may be explained by the overprescription of antibiotics, patients not finishing their complete antibiotic therapy after symptoms stop persisting, and by the subsequent natural selection of bacteria. When an infected person takes antibiotics, it is the relatively weak bacteria that perish first. The stronger bacteria linger and proliferate, sometimes (though more and more frequently) mutating and becoming resistant to antibiotics.

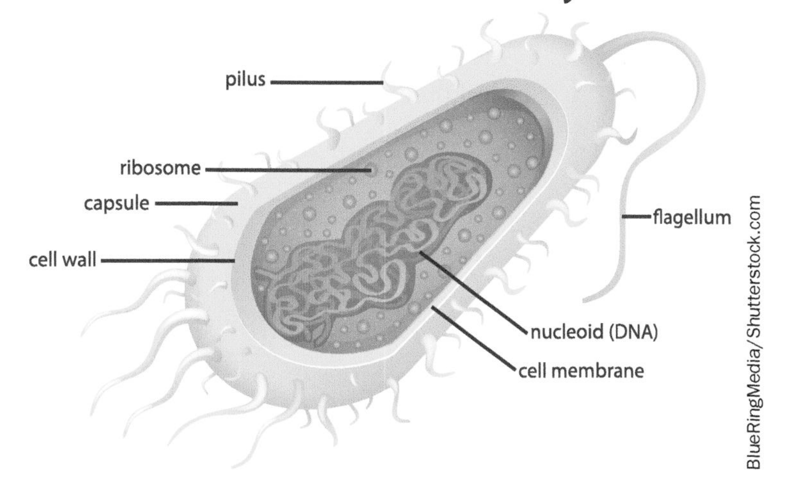

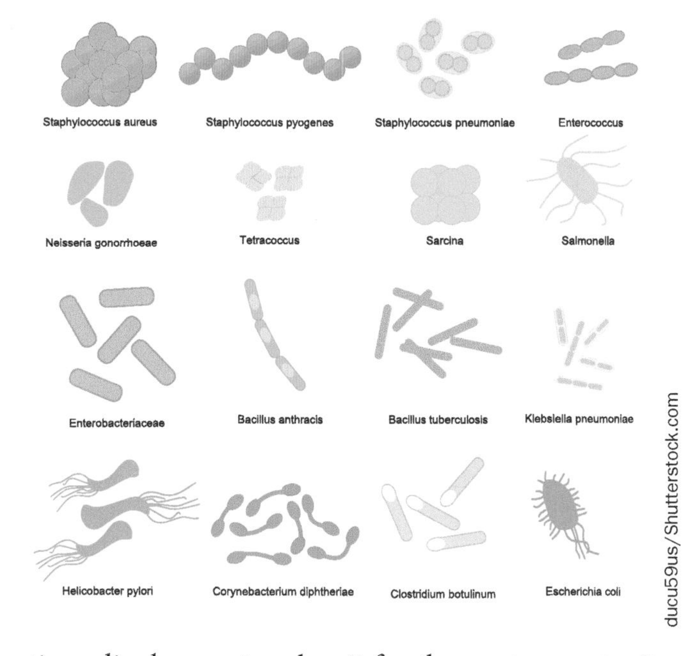

These bacteria oftentimes lie dormant and wait for the next opportunity to infect a host—an infection sure to be worse than the last.

The bubonic plague that wreaked havoc on 14th century Europe was caused by bacteria found inhabiting the guts of fleas. When ships would come to port in medieval Europe, rats would come down the ropes into port and the fleas on a rat would bite an unsuspecting victim and then the bacteria would infect the lymph nodes, spreading the disease in the victim while simultaneously disabling the immune system. Swellings of the lymph are called **buboes**, which explain the term *bubonic plague*. Present day, these bacteria are treated with antibiotics, and instances of plague are rare.

Bacteria get a bad reputation because they are so often pathogenic; but, in fact, we need bacteria. If we were to compare our own DNA to the DNA of bacteria that lives symbiotically with our bodies, we have roughly 90% bacterial DNA. Bacteria form a symbiotic, mutualistic relationship with humans to assist with our digestion and to help outcompete pathogenic or disease-causing bacteria. Collectively, our good bacteria are called our natural flora. Some people eat yogurt or take probiotics to replenish their natural flora.

Fun Fact

Humans have 10 times more bacteria than cells in our body; therefore, if we didn't have bacteria, we would be hazy looking versions of ourselves![iii]

Viruses

Viruses exist somewhere between alive and not alive. Because they do not possess all the emergent properties of life such as autonomic metabolism and reproduction (see Chapter 2), they are technically not considered alive. A virus consists of a protein coat surrounding nucleic acids, typically **DNA** and, sometimes, **RNA**.

A virus requires a host in order to replicate and metabolize food. Their protein coats vary in size and shape and are instrumental in classifying a virus.

Viruses are also categorized by the type of host that they infect, which can be bacteria, plants, or animals. Viruses usually stay within their host range but are known to mutate and infect a new type of host. Like bacteria, they are not always pathogenic. In fact, they often infect a host cell and remain dormant, showing no symptoms at all.

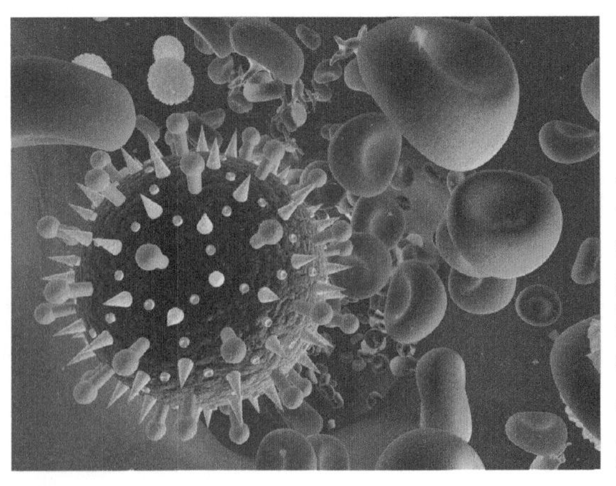

Sebastian Kaulitzk/Shutterstock.com

Influenza Virus

Small pox, chicken pox, polio, and HIV (the virus associated with the disease AIDS) are examples of pathogenic viruses. HIV is a **retrovirus** that means it creates DNA from RNA. Most viruses do the reverse: they create RNA from DNA. The retrovirus incorporates itself into the host genome and never leaves. Viruses that infect a wide host range tend to be the most harmful. The West Nile virus has a wide host range that includes birds, mosquitos, horses, and people. Some viruses, such as measles, only infect people.

Have you ever noticed that many viral infections, such as avian flu, SARS, swine flu, and HIV, are extremely virulent and deadly in their onset? These viruses are considered **emerging viruses** because they seem to come out of nowhere and inflict tremendous harm and illness. Mutation rates are high as these viruses jump from host to host, becoming very infectious and virulent. But over time, virulence begins to wane. Knowledge regarding curbing the infection rates is spread and people learn to protect themselves, changing their habits to prevent infection or to keep it from spreading. Also, from a natural selection perspective, it is an evolutionary dead end to kill the host, so it is in a pathogen's best interest to be as tolerable as possible, and typically will become so over significant evolutionary time.

Smallpox is a virus that has been around for centuries. In fact, it decimated a huge percent of the native American population, who were isolated from the common European virus. A virus called vaccinia or cowpox, which is found in cows, is similar to smallpox. Milk maids and persons who spend considerable amounts of time around cows appeared to be immune to the closely related smallpox virus. What would explain their immunity?

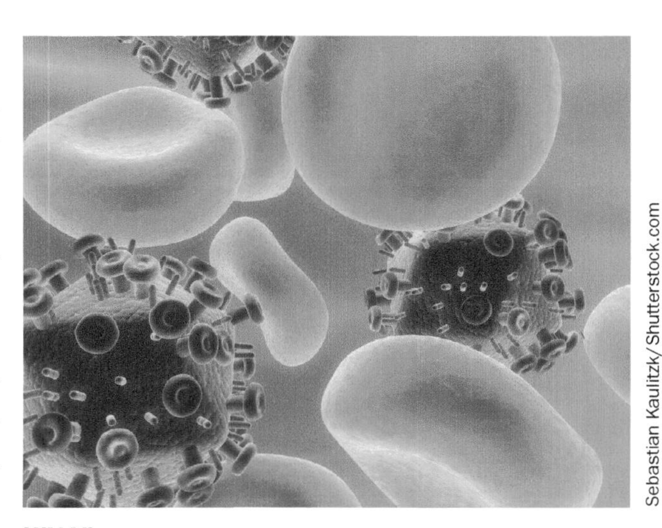

Sebastian Kaulitzk/Shutterstock.com

HIV Virus

The Milkmaids' Immunity

Our immune system works so that, if we are exposed to a pathogen, our body stores information about that pathogen (in this case, the virus). When it sees the pathogen again, it recognizes that it is potentially harmful and initiates the appropriate immune response. This is considered **adaptive immunity**, meaning that our body adapts and responds specifically to pathogens that we recognize because of previous exposure.

Lady Mary Montagu (1689–1762) was an English woman who paved the way for the smallpox vaccination. The vaccine was developed after it was noted that people exposed to the cowpox virus or a weaker strain did not get sick. Lady Montagu used powder from wounds of a less virulent strand of the virus. The cowpox virus was close enough in structure and genetic profile to the smallpox virus that the immune system would recognize it as being harmful and the body would generate the appropriate antibodies.

The word vaccine comes from the word cowpox, vaccinia. Native Americans were tragically vulnerable to the virus, in part, because they did not have a legacy of living among livestock as westerners did, and their immune system did not evolve accordingly.

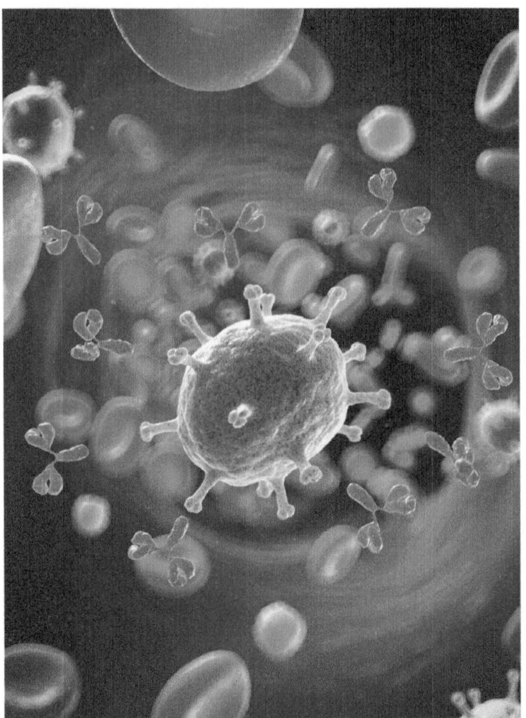

Sebastian Kaulitzk/Shutterstock.com

Antibodies Attacking a Virus

As do all microbes that are pathogenic, viruses tend to have a negative connotation. But, like bacteria, viruses have many beneficial functions. Some viruses can kill bacteria that are harmful to humans. Also, they assist in providing variability to many organisms when viruses infect the host with benign genetic information. But if it occurs on an important genetic site, it can potentially cause cancer in the host. It is believed that viruses have helped humans evolve by integrating their genetic information into humans and adding variability to the human genome.

Fungus

Thirty percent of all known funguses are parasites—but not just to humans and other animals—they are primarily parasites and pests to agriculture. Fungi can have airborne spores found 160 km above the ground! They feed on dead organic matter and digest nutrients outside of the fungal structure, recycling them back into the ecosystem. **Mycosis** is a general term for a fungal infection in humans. Examples of parasitic fungi include athlete's foot, ringworm, yeast infections, and candidiasis.

The black mold *Aspergillus* is toxic and known to be carcinogenic.

Fungi also assist plants in obtaining nutrients from the soil and are instrumental to ecosystem health. Mychorrizal fungi form a mutualistic symbiotic relationship with plant roots and

Damian Herde/Shutterstock.com

A Fungal Stem Disease

help absorb nutrients from the soil by nitrogen fixation. They have also been instrumental in assisting angiosperms colonize land.

Protozoa

Protozoa are small unicellular eukaryotic organisms. They are larger and more complex than bacteria and, unlike bacteria, contain a nucleus. Protozoa are responsible for the African sleeping sickness and can also cause dysentery which can affect all vertebrates that drink fecal contaminated food or drinking water. Amoebic Dysentery affects 40 million people per year.

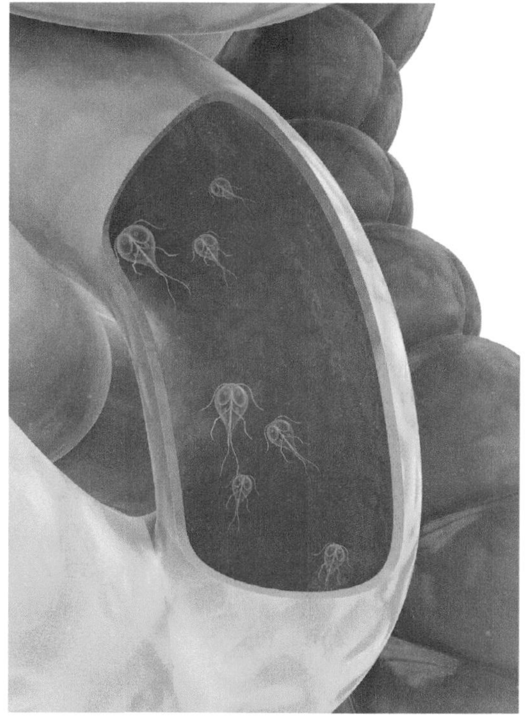
Sebastian Kaulitzk/Shutterstock.com

The Protozoa That Causes Dysentery

The microbes that are responsible for the disease malaria are also protozoa; they infect red blood cells. However, there is a genetic mechanism that can endow resistance to malaria—at a price. But first, it is important to understand genetic zygosity. People who have genes from both parents for a particular trait are said to be **homozygous** and carry two of the same active genetic alleles, while people who carry a single active genetic allele for a particular trait from one parent only, are considered **heterozygous**.

Persons homozygous for normal red blood cells are vulnerable to malaria. On the other hand, persons homozygous for the disease sickle-cell anemia (in which red blood cells rupture or lose their shape and thus cannot effectively carry oxygen) are vulnerable to the disease and face the possibility of an early death. A person heterozygous for sickle-cell anemia is resistant to malaria due to high red blood cell turnover and can live normal lives, provided they are careful at high altitudes where oxygen is scarce. This suggests that sometimes it is evolutionarily advantageous to be a heterozygous carrier for a genetic disease—a reminder that evolution is based on chance.

Prions

Prions are proteins found in the central nervous system of both humans and animals. For reasons still unknown, prions can become infectious, circulating in the central nervous system and denaturing normal proteins (causing them to lose their shape). These denatured proteins can, in turn, become new prions and continue the propagative and usually exponential genesis of new prions. As a result, the nervous system does not work properly and death is imminent. Its causes are linked to eating beef or carnivorously fed cows that are infected with preexisting prions.

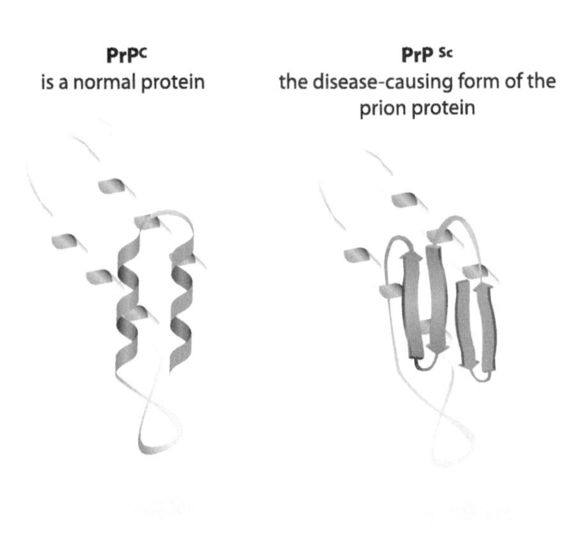

Designua/Shutterstock.com

Cancer

Mutagens are any agents that increase the frequency of mutations to DNA. They range from viruses to radiation. Mutagens are often carcinogens. A **carcinogen** is a cancer-causing agent. Carcinogens are anything that causes cells to mutate and change from having normal, healthy cell growth to prolific unhealthy, unregulated growth—eventually creating a tumor (ball of cells). Examples of carcinogenic agents include tobacco smoke, asbestos, radiation, viruses, and black mold.

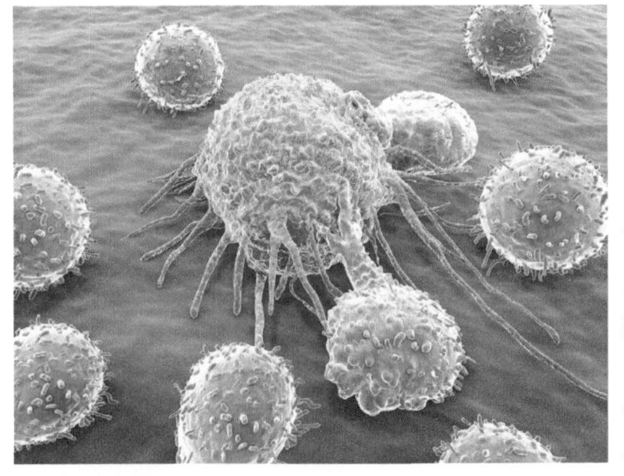

Juan Gaertner/Shutterstock.com

Cancer Cell and Lymphocytes

Cancer is almost always a nontransmittable disease; it cannot be spread from host to host. It develops from genetic origins, from environmental exposure, or a combination of both. One exception is the cancer afflicting the Tasmanian devil, which are currently on the brink of extinction due to a contagious type of facial cancer with a 100% mortality rate. It spreads from one host to another through biting. Scientists are desperately trying to come up with a vaccine before they become extinct.[iv]

Healthy, noncancerous cells typically have the following characteristics: **Anchorage dependence**—healthy dividing cells stay anchored to a substrate (its place of origin) and do not invade other tissue; **contact inhibition**—when healthy cells divide, if they come into contact with too many cells, they innately know to stop dividing. This may be likened to pressing an off switch that stops cells from dividing. Healthy cells typically divide 50–70 times, then perform **apoptosis**—a type of cell suicide. This keeps entropy (disorder) from occurring in healthy cells (the older the cells become, the more mutations they are likely to have accumulated) and it also maintains a healthy number of cells.

Cancerous cells, on the other hand, typically do not have the above health-inducing characteristics. Rather, cancer cells tend to be detrimental to organisms; they invade other tissues and perform **angiogenesis**—the creation of vascular tissue to directly feed tumors. Tumors do not recognize their naturally termination sequences—something regulated by DNA telomeres.

Telomeres are the noncoding ends of DNA. When telomeres shorten, it signals the cell to commit apoptosis or cell suicide. Through its telomeres, it is almost as though DNA is aware of how old it is. Consider, for example, the young, cloned sheep Dolly that made national headlines a few years ago; she had problems associated with a middle-aged sheep such as arthritis because she was cloned and therefore born with shortened telomeres (the clone cells maintained the same age as the original Dolly). Telomerase is an enzyme that restores the length of telomeres—the ends of DNA. Therefore, the cells never perform apoptosis.

If a tumor (a ball of cells) stays within in its place of origin and does not have the ability to invade neighboring tissues or move around the body, it remains benign. On the other hand, if it starts to invade other tissues, it is considered malignant. If cancerous cells get into the lymph or blood stream they have access to the rest of the body; this is when the cancer is considered to have **metastasized**. Names for the different types of cancer are associated with the types of tissue on which they originate: **sarcomas** form in muscle and connective tissue; **carcinomas** originate in epithelial tissue (skin); and **leukemia** originates in blood and lymph.

TELOMERASE

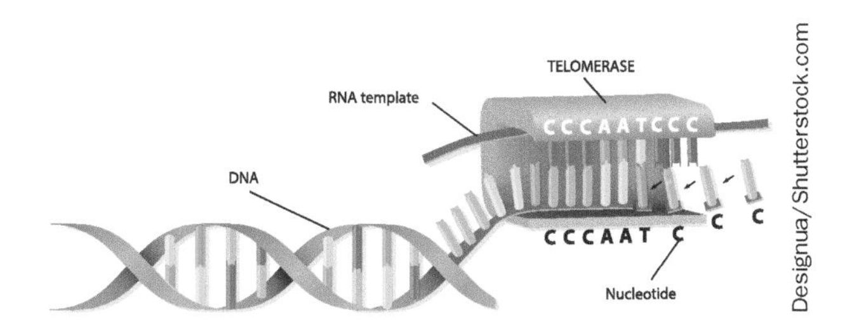

Designua/Shutterstock.com

Radiation treatment involves treating the tumor with intense focused energy to kill the cancer. Patients typically go for anywhere between 6 and 12 rounds of treatment. Prolonged radiation can compromise the integrity of bone marrow, and in extreme cases, a bone marrow transplant is needed. **Chemotherapy** is the process of circulating chemicals throughout the body that target dividing cells. Cells that give rise to hair are constantly dividing which explains hair loss in chemo patients.

Fun Fact

The naked mole rat is the only known organism that scientists cannot successfully induce with cancer. No known mole rat has cancer. This is primarily due to an extremely conservative regulatory protein that causes cells to apoptose if *any* crowding occurs, forcing proliferation to be virtually impossible.

Other Environmental Risks

Teratogens are anything that can cause birth defects by affecting fetal development. They range from chemicals in the environment, to deficiencies that are linked to birth defects. Sometimes teratogens stem from a combination of factors. Examples may include tobacco smoke, radiation, viruses (such as herpes), and even trauma during fetal development.

Corrosive agents are reactive chemicals that damage or compromise the surface of a material. Sulfuric acid and battery acid are examples of corrosive substances.

Combustible items have a strong capability of igniting or burning at normal temperatures. Ethanol, solvents, and paint thinners are often combustible.

Toxic chemicals may cause either short- or long-term harm depending on the frequency and duration of exposure, and the chemical's damaging properties. Examples of toxic chemicals are lead, mercury, and asbestos.

HAAs (hormonally active agents) are environmental contaminants that act as endocrine disruptors. They are often chemically similar to hormones such as estrogen which can trick the body into responding as it would to the natural hormone. Some pesticides, fire retardants, and common consumer goods that contain phthalates (see Chapter 8) are HAAs. Exposure of humans and wildlife to high concentration to HAAs can cause disruption of development, anomalies in behavior, and irregularities in the immune system.[v]

How Our Body Protects Itself from Pollution

Our respiratory system is equipped with **mucus**—a thick, sticky substance found in the nose and other body cavities, whose job is to trap and flush out dust and other impurities.

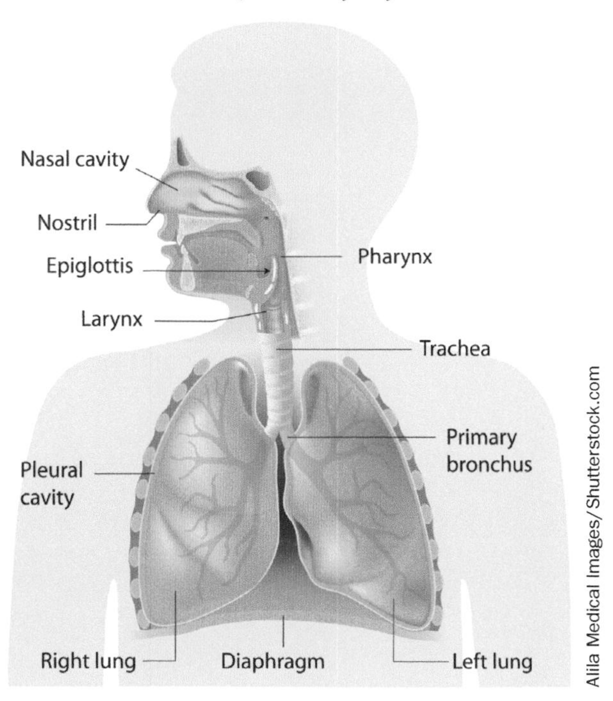

Our respiratory system is also equipped with **cilia**—small hair-like projections whose function is to sweep out irritants (in an upwards motion) and impurities, thus protecting our lungs. But when the respiratory system is overwhelmed due to excessive pollution, the cilia become brittle and stop functioning properly. Impurities build up in the respiratory system causing a possible infection in the bronchioles and, in some cases, the lungs. This leads to bronchitis—and in more extreme cases, can lead to emphysema, asthma, chronic obstructive pulmonary disease (COPD), similar to what long-term smokers experience—and in rare instances, mesothelioma or lung cancer.

Review Questions

1. What is a "pathogen?" What are common pathogens?

2. Compare and contrast healthy cells with cancerous cells.

3. Why is some immunity considered "adaptive?"

4. What is an "emerging virus?"

Discussion Questions for Class

A. Why was Dolly, the cloned sheep, born with health problems typical of an older sheep?

B. Can you think of examples of infectious diseases? What about nontransmittable diseases?

C. Why are carcinomas among the most common types of cancer?

D. What lifestyle changes can you make to reduce your risks of cancer?

Your Personal Sustainability Journal

Buy rubbing alcohol from a drug store. Put it in a spritzer bottle and use the alcohol as a disinfectant to clean keyboards, cell phones, and other common household items.

Get creative! Do a web search and discover how to create your own toothpaste using baking soda and peppermint oil.

Make your own natural deodorant, or make a point of using deodorant that does not contain aluminum.

Cut down on shampooing, or consider not shampooing your hair at all! It is very common in most parts of the world to shampoo once a week. Many people are opting to not shampoo at all, but are using baking soda and apple cider vinegar to wash hair occasionally. This will restore your hair's natural oil, and you will not be exposed to chemicals (sulfates, parabens, petro chemicals) found in shampoo. The chemicals will not enter the watershed, and you will reduce you reliance on plastic bottles.

If you have prescribed antibiotics, take them all. Do not stop taking them just because you feel better. Listen to the advice of your health professional.

Don't smoke!!

Keep your immune system healthy by eating healthy and living a healthy lifestyle.

Exercise and meditate; engage in such activities that reduce stress levels. Make it a priority.

Add your own suggestions here.

Endnotes

i. http://www.hhmi.org/research/cell-cell-communication-bacteria

ii. University of Wisconsin-Madison (2013, May 23). Chemists find new compounds to curb staph infection. ScienceDaily. Retrieved January 19, 2014, from http://www.sciencedaily.com/releases/2013/05/130523093321.htm

iii. http://www.scientificamerican.com/article/strange-but-true-humans-carry-more-bacterial-cells-than-human-ones/

iv. Hannah V. Siddle et al. Reversible epigenetic down-regulation of MHC molecules by devil facial tumour disease illustrates immune escape by a contagious cancer. PNAS, 2013 DOI: 10.1073/pnas.1219920110

v. National Research Council. Hormonally Active Agents in the Environment. Washington, DC: The National Academies Press, 1999

Chapter 15

A Sustainable Planet

"Green" is all about providing conditions that keep our environment and our quality of life healthy; therefore, any green discussion would be incomplete without addressing the **social**, **working**, and **political** conditions that affect our environment and our lives.

This chapter discusses the green state of our social and working environments, and also the politics that affect how the world is coping with Mother Earth's trials and tribulations.

DID YOU KNOW?

Until the 1960s, there was not considerable momentum for wildlife and biodiversity preservation, at least it was not the nexus of global concern that it is today. Dialogue and environmental concern slowly gained traction during the 1960s as human population growth reached its biotic potential and inevitably placed strain on natural resources. With enough people understanding that the state of the natural environment is closely entwined with the quality of human life, preservation efforts expanded on both local and global scales.

In 1964, the ***National Wildlife Preservation System*** in the U.S. was implemented, protecting Federally managed land and working with the Federal Agencies (such as the National Parks and U.S. Wildlife Fish and Game) that manage them. In 1969, ***NEPA***—an act by Congress to improve environmental quality—was passed. In 1970, the ***EPA*** (Environmental Protection Agency)—the Federal Organization through which all Environmental Legislature moves—was created. In 1971, ***Greenpeace*** was established. Starting as an antiwar protest, Greenpeace is comprised of a group of volunteers who consider themselves peaceful warriors fighting to protect the environment. Nonviolently, they also expose those who commit offenses against the natural world. In 1972, ***UNEP*** (United Nations Environment Programme) was created. UNEP is the UN's advocate for stewardship of the environment, promoting education and sustainability. In 1972, the ***Noise Control Act*** was passed. Enforced on a national level, it serves to protect the health and well-being of people, wildlife and the environment by promoting a noise-free environment. In 1973, ***The Endangered Species Act*** was passed, protecting species that are about to go extinct and their habitats. 1973 ***CITES*** (The Convention on the International trade in Endangered Species of Fauna and Flora) was created. It is an international treaty protecting wildlife from poaching and ensuring their well-being during international sales and trade. ***The Fishery Conservation and Management Act*** of 1975 protects the U.S. coast from foreign fishing—up to 200 miles out. In 1980, ***The Alaska National Interest Lands Conservation Act*** was passed—designating large parts of Alaska as National Forests, Parks, and Wildlife Refuges. 1980s and beyond brought us of an ivory ban, a whaling ban, the ***Montreal Protocol*** (to protect the ozone layer), and ***Kyoto Protocol*** (to reduce greenhouse gas emissions). The collective determination and gestalt effort of these organizations has brought environmental and wildlife concerns to the forefront of global awareness.

Many of our major environmental problems can be traced back to overpopulation—and with good reason! As of 2019, there are over 7.5 billion people (according to census.gov). This number is staggering considering that there were only just more than 1.5 billion people on the planet in the year 1900. This considerable growth in such a short time exhausts natural resources that do not have time to regenerate before they are repeatedly harvested. It also generates waste and pollution that causes many environmental, ecological, and health problems. (We will explore these in Chapter 8.)

A Growing Human Population

Consider that before the agricultural revolution 12,000 years ago, there were 5 million people on the planet. Present day, there are over 7.5 billion. What may account for this abrupt population growth? There is no simple answer to this question, but we can postulate that exponential human population growth can perhaps be attributed to several salient factors: the availability of new habitats that have become readily available through technology and exploration; the ability to feed many people through industrialized farming; and advanced, wide-spread health care and sanitation that reduces infant mortality and overall death rate.

Let us also examine our recent history and observe seminal events that may have led to such a population spike.

Before the glaciers receded, there was no evidence of farming among *Homo sapiens* (what is now modern humans). Our lifestyle at this time would be characterized as that of a hunter gatherer, and we were environmentally constrained by resources that were immediately available to us.

Approximately 12,000 years ago, this all changed as we started cultivating plants and raising livestock. Animals such as sheep, pigs, and cows became domesticated as we started cultivating barley, wheat, rice, and peas for crops. As this systematic agriculture practice came into full swing, it is possible that we contemplated a practical use for our surplus of crops, lest they go to waste. Such food extras, having value, may have been stored and bartered for, increasing our socialization.

Soon thereafter, we started settling into villages and cities. We also became less constrained by the environment as food and resources became stored, shipped, and exported. Because we were no longer limited by the environment and what was only immediately accessible to us, we started consuming in excess—and what we now consider unsustainable practice started becoming common.

Not having to hunt and gather for our food meant we were freed up to start compartmentalizing our skill set, perhaps setting the stage for job specialization, entertainment, and individual roles in society. This increase in human interaction ripened environmental conditions for a dramatic increase in human population growth we observe present day.

Demands of a Growing Population

Dramatic increases in the human population began with the onset of the Industrial Revolution over 200 years ago. This growth trajectory was a cause of concern to economists, scientists, and demographers who had never observed such an unprecedented acceleration in human population growth. An economist named Thomas Malthus, interested in the relationship between food supply and population growth, wrote *An Essay on the Principle of Population.* This best seller discussed how populations grow exponentially, that is, steadily and unchecked, while natural resources are limited, ultimately causing a struggle among individuals and groups for food and resources. Malthus's writings would later be instrumental in assisting Charles Darwin to formulate the theory of evolution through natural selection.

Unchecked human population growth can have many undesirable side effects including poverty, pestilence, hunger and poverty, droughts, exploitation by human investors, violence, and malnutrition. A growing human population places a high demand for construction of buildings and other infrastructure—schools, roads, and hospitals, etc. It also predicts demands on a country for water, electricity, and food—all of which encroach upon species habitats, driving them out and, possibly, toward extinction. Overdevelopment of land leads to soil runoff that strips nutrients from the soil causing **desertification**. Industrial agriculture, a practice that effectively feeds many people, also leads to desertification.

Additionally, oil, coal, and other biomass are burned to meet the energy needs of the growing population, accelerating the amount of carbon released into the atmosphere and leading to changes in global temperature as well as increases in the amount of acid rain (see Chapter 2). In the year 1958 (during the baby boom that occurred between 1946 and 1964), the human population growth rate was the highest ever recorded. The good news is that the rate of population growth is slowing down, such that the slope of the growth curve has decelerated. According to the United Nations Population Division, from 2015 to 2020 the population growth rate of developed countries is 0.26%, with a rate of increase at .6%. Compare that to the U.S. rate of increase at 4.3% and the U.S. rate of population change from 2015 to 2020 is about .70%. According to the CIA's World Factbook, more than half of the world's countries are now below the replacement rate of 2.1%. **TFR** or total fertility rate is the amount of children expected to be delivered by women during their reproductive years. According to the UN, the global TFR from 2015 to 2020 is 2.47. More developed countries have a TFR of 1.69 while less developed countries have a TFR of 2.59. The U.S. has a TFR of 1.89, just a little above other MDCs.[i]

World population growth rate may not seem like a significant increase but given the already large population, even a small amount of growth makes a huge impact—in terms of pressure on the earth to provide for and sustain such growth.

Globally from 1955-1960 the TFR was 4.89. The United States was 3.58 for this time period. From 2015-2020 the median age for childbearing in MDCs was 30 and the median age for childbearing in LDCs was 28. What accounts for this shift? Factors that affect birth rate include the cost of raising children, education and employment for women, mortality rate of infants, necessity of children in the work place, availability of retirement, average age of marriage, and access to safe abortions or birth control, to name a few. There is an observable negative correlation between the education level of a woman and the number of children she is expected to have in the course of her lifetime. So, one might conclude that in the developing nations where women have little access to education and job opportunities, women of reproductive age are more likely to have more children. In fact, some developing nations experience population growth rates as high as 4% a year! Tanzania grows at 3.15% per year as of 2020.

Developed nations have a stabilized economy and are considered *industrialized*. Citizens of developed nations as a whole have better access to education and health care. Developed nations are also called **MDCs**, an acronym for "more developed country." The economy of **developing nations** is not yet stabilized, industrialization is a work in progress, and, as a whole, its citizens do not have ready access to higher education. Developing nations are also called **LDCs**, an acronym for "less developed countries."

Given all of the factors that influence birth rate, it should come as no surprise that population growth rate in developed nations is significantly slower than that of developing nations. Currently, the population growth rate of some developed nations is below the replacement rate—A negative growth rate suggests that the number of individuals that emigrate out of the country or die *exceeds* the number of those who are born in that country. War and violence can also account for a negative population growth rate. What do you suppose accounts for such negative growth trend?

The exception to the trend among developing nations is the United States which, as of 2011, experienced a growth rate commensurate with the growth rate of developing nations. This may be explained by the high volume of immigration into the United States.

BOTH death rate and birth rate influence population size. The industrial revolution brought with it not only exponential population growth but also modern health care, medicine, and sanitation—all instrumental in decreasing the death rate worldwide. This is reflected in the changing trend of both global and regional death rates. Consider the fact that some historians maintain that the worldwide life expectancy in the year 1800 was as low as 28! This is explained through poor health care and lack of modern medicine, causing individuals to die from infections readily treated with antibiotics today. In the year 1900, the life expectancy of the typical American was 47 years. As of the year 2000, the life expectancy in the United States was 77 years and continues to climb.

Due to exceptionally low death rates, individuals in Germany. As of 2018, Germany, Israel and Italy enjoy a life expectancy above 80 years old. Japan has a life expectancy of over 85 and Monaco has a life expectancy of almost 90!(Source CIA World Factbook). Afghanistan have a life expectancy of 52 and Zambia 53 (according to CIA World Factbook). Due to high death rates, relatively few individuals survive beyond the age of 35 living in such countries as Nigeria and Guatemala. Countries with high percentages of young people grow rapidly because they have such a high number of individuals of reproductive age. Countries with high percentages of older individuals (above reproductive age), on the other hand, grow more slowly.

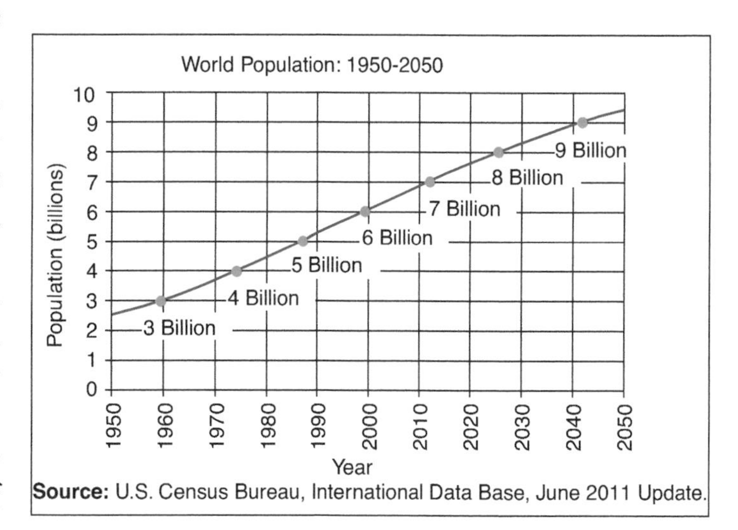

Source: U.S. Census Bureau, International Data Base, June 2011 Update.

Life span and death rate, together, offer invaluable information about quality of life. The human population in Japan has a life expectancy of around 80 years. This high life expectancy indicates a healthy lifestyle, healthy diet, and quality health care. On the other hand, nations with low life expectancy, such as Afghanistan, indicate poor health care and high rates of disease, famine, and violence.

Urbanization

In the year 2007, for the first time ever, there were a greater percentage of people living in urban areas than in rural areas. Consider that in the year 1950, roughly 30% of the world's population lived in urban areas. As of 2014, 54% of the world's population lived in urban areas. By 2050, it is estimated that 66% of the world' population will be living in urban areas, cities and metropolises.

The world's most urbanized countries are North America, South America, Europe and the Caribbean. Roughly 90% of the world's most rural areas are in Asia and Africa, but these areas are also urbanizing quite rapidly.

Tokyo is the world's most populated city with 38 million people (as of 2014). The second most populated city is Delhi with 25 million people, followed by Mumbai, Mexico City, and Sao Paolo, each with roughly 25 million people.

From 2010 to 2015, MDCs had an urbanization rate of .3%, while LDCs had an urbanization rate of 2.9%. From this data, we see that less developed nations are urbanizing at a much greater rate than more developed countries. (As of 2015, MDCs contain 1.3 billion people, projected to be 1.3 billion in the year 2030. LDCs contain 6.9 billion people as of 2015, but are projected to have 8.3 billion people as of 2030).

With urbanization increasing, and many people moving to urban areas, there are questions and concerns regarding such growth. In particular, there are three factors that place high stress on sustainable urban development, including *environmental protection, social development,* and economic development. These factors are all considered with consistent and predictable growth. Unfortunately, rapid and unplanned urbanization makes sustainable development in all of these areas a big challenge.

Consider that urbanization is often linked to higher literacy rates; longer life expectancies; lower fertility; and more mobility throughout various geographic areas. It also typically means better health services, better public transportation, and more participation in political and cultural endeavors. However, unplanned urban expansion often leads to pollution, environmental degradation, as well as unsustainable production and consumption of goods.[ii]

Can Microcredit Help Curb Population Growth and Eradicate Poverty?

Let us recall that education and employment of women are both factors that decrease birth rate. Women who are employed and financially independent tend to give birth to fewer children and therefore have a greater chance of breaking the cycle of poverty. **Microcredit** is a system that involves lending money to the poor, and, in many cases, mostly women, who have no collateral in many areas of the world. This loan is usually a couple of hundred dollars, suitable or commensurate with the need of recipients which is usually modest relative to U.S. standards. The loan is enough to help get a business started and generate income, but not an overwhelming amount so loan recipients cannot get out of debt. The goal is to help recipients become financially secure, breaking the cycle of poverty, and potentially strengthen the economy and stabilize population growth.

Microlenders also offer financial assistance and knowledge on how to get a small business started and generate income, ensuring that payback rates are high. There are several microlending banks including BancoSol in Bolivia and Women's World Banking (WWB) which only offers financial assistance to women.

Dr. Muhammed Yunus, an economist from Bangladesh who spent several years teaching and studying in the United States, returned home to his native land only to find the people impoverished. Wanting to help his people get out of debt, his efforts included microlending and soon materialized into the Grameen Bank in 1983. Its goal is both social and economic, to help people living in extreme poverty become financially independent. This practice of microlending set a precedent for the international community and became a more global endeavor. (Recognized for his selfless actions, Dr. Muhammed Yunus and the Grameen Bank won a Nobel Peace Prize in 2006. [iii]) The Grameen Bank aspires to help achieve the first Millenium Development Goal, to eradicate extreme poverty (MDG 1). Please see Chapter 15 for more about the Millennium Development Goals.[iv]

lev radin/Shutterstock.com

Nobel Laureate Muhammed Yunus

Work and Environmental Health

OSHA

Nearly 13 work-related deaths occur daily in the United States! This may seem like a staggering number, but the rate has significantly declined from 38 worker deaths per day in 1970!

What was the impetus for such a dramatic reduction in worker deaths? In 1970, under the Nixon Administration, the Occupational Safety and Health Act of 1970 was passed, and from it, **OSHA** was created under the Department of Labor.

OSHA (the Occupational Safety and Health Administration) was created to "assure safe and healthful working conditions for working men and women by setting and enforcing standards and by providing training, outreach, education, and assistance." The creation of OSHA dramatically altered the fabric of the workplace environment and placed worker safety as its main objective.

One key element crucial to OSHAs success has been improving overall workplace safety standards by demanding employers control the spread of pathogens and provide workers with proper protection from them. It ensures worker safety against hazardous material and provides worker protection from noise and respiratory irritants. It also regulates and monitors the use of heavy machinery and equipment that may cause accidents.

OSHA has jurisdiction over private sector employees and conducts on-site inspections periodically. Employers are required to comply with OSHA standards (www.osha.gov).

NEPA

Workplace health, safety, and quality were not the only environmental concern in the year 1970. As the environment continued becoming polluted and its natural resources started waning, the public started to become increasingly concerned about the amount and quality of natural resources—what would happen to our environment and natural resources if we continued to destroy and pollute them at the going rate?

As a response to this growing concern, Congress passed the National Environmental Policy Act of 1969 (**NEPA**) designed to protect and maintain overall environmental quality. Although NEPA protects the environment, NEPA also recognizes the social and economic need for natural resources and seeks a balance between human need and overall environmental quality. NEPA assesses the overall environmental impact of human activity and tries to either prevent or mitigate its damage. It also implemented **CEQ** (the Council of Environmental Quality) to oversee its execution.

Prior to any large construction project or human activity that may alter the environment or natural resources, CEQ performs environmental assessments, **EAs**. EAs examine if such construction or human activity would have potential negative impacts on the environment and overall human quality of life. If the assessment is affirmative, an environmental impact statement (**EIS**) is prepared. EISs examine and analyze all possible issues related to the project, particularly, how it will impact both the environment and the quality of human life. The report is then prepared for public review. EISs also propose alternatives if impact is significant.

Economics and the Environment

Many people feel that the next major shift involves our perspective on economics. Many feel that if we want to conserve natural resources, live more sustainably, and have a more socially inclusive global economy, we must *shift* our rigid ideas about the economy, as well as alter our economic models. Let us explore some economic philosophies and their view on the environment:

> **Classical Economics** gained widespread popularity in the 1800s—This ideology is based on free trade, competition, and a free and open economy. It is predicated on the notion that everyone is able to pursue their interests and produce their product at will. It espouses no government restrictions or interference, and an economy built on capital (clothes, cars, appliances, etc.). The main objective of classical economics is to increase goods and services and the **GDP** (Gross Domestic Product—the value of goods and services produces annually by a given country.) Classical economics relies on constant growth, since growth keeps people employed, avoiding the cycle of poverty, and therefore keeping them happy. It also sees material goods as interchangeable—when one runs out, it can be replaced with another. However, from an environmental perspective, classical Economics does not recognize limitations of natural resources and limits to growth, nor does it recognize the finite demand for those goods.

Environmental Economics is a subdivision of economics, applying the environment to its economic model. Environmental economics supports the notion that many of our natural resources come from the environment and that we can use them to build an economy with its capital, but we must acknowledge its limitations and use it sustainably.

The current beliefs about our economy are mainly concerned with extraction of natural resources and privatizing and ultimately selling them for profit. Critics are concerned that logic surrounding the relationship between our natural resources and the economy are flawed. As it exists now, the derivatives of our ecosystem are worth more monetarily than an intact, fully functioning ecosystem with all its component parts. This flaw in logic, many argue, places economic value ahead of intrinsic and ecological value. This, they believe, is unsustainable and will inevitable cause our demise.

Many environmentalists say that perhaps it is time we challenge our ideas about how we define an economy, as our economy often systematically treats natural resources as though they are mechanized and fungible, rendering its supply to be unsustainable. Exploiting our natural resources in this manner implies a separation between society and natural resources. Some argue that it is time for this discrete separation to be challenged. Instead of a materials economy, perhaps it is time to invest in an economy that is based on science, innovation, culture, art, social inclusion, health, and meaningful interactions, allowing the natural world to be treated as invaluable.

This brings us to **Ecological Economics**—which focuses on the true value of natural capital as well as the best quality of life for humans while preserving the natural world. It explores our complex interaction with the natural world, aiming to ensure fair distribution of natural resources and sustainable natural capital. It supports social justice and justice for the environment. It views these components as one cooperating system.

This approach to economics considers the long-term consequences and so is as nonconsumptive as possible. It is a multidisciplinary means to connect ecology and the environment, incorporating human health, education, and meaningful social interaction—all factors that contribute to a thriving and prosperous society. It supports the notion that nature is the foundation of economic activity, acknowledging and optimizing humanity's place in the ecological world. It also pursues sustainable extraction as well as fair and efficient distribution.[i]

Co-Ops

Another option for building a long-term sustainable business and economy is to support or start your own worker cooperatives (co-ops). Worker co-ops are unique in that they are controlled and owned by all of its employees. All major business decisions must be agreed upon by all of its workers and on every level of operation. If a business wants to relocate to another city, for example, then every employee must agree to this decision.

The benefits of worker co-ops are many. Areas that have many co-ops typically have high employment rates. This helps break the cycle of poverty (a major cause for environmental problems). Workers build their entrepreneurial spirit and acquire the skills necessary to run a business. This system fosters dialogue and democratic thinking, since all employees must learn to compromise and work together. It is an engine for social change and social justice, especially in poor and disenfranchised areas.

Another benefit of worker co-ops is that they often have close ties to the people of their home city or town. This makes it less likely for the business to engage in environmentally irresponsible practices, which would otherwise alienate potential clients and associates. So in this sense, worker co-ops are also environmentally sustainable.

Are We Moving Toward a Global Green Economy?

The United Nations Environment Programme's (UNEP) Green Economy Initiative is a global endeavor promoting a green economy through education and advisement to the participating countries. Green in this context is defined as low carbon, resource efficient, and socially inclusive.[ii] A green economy, in other words, is one that builds jobs and promotes income based on social equity, sustainable use of resources, low pollution, the eradication of poverty, and high energy efficiency. This initiative relies on rigorous economic research and strong policies reflecting and supporting such research.

The initiative crystallizes the notion that a green economy at this point is not only beneficial but also *essential* to decreasing poverty and increasing social and economic growth. It puts forth great effort to preserve or enhance natural resources and natural capital when possible. This is imperative for people, especially, in poor nations that are rich in natural resources but low in income and employment. The program instructs such countries how to use their biodiversity to build a strong and sustainable economy. Indeed, this effort has been met with marked success. Current research indicates that a green economy creates green jobs and reduces poverty in countries such as South Africa, Nepal, Peru, and Namibia.

The annual Green Economy Report is written by the UNEP's Green Economy Initiative. It provides evidence that the greening of economies does not slow them, but rather enhances them. The report examines a multitude of topics including, but not limited to: efficient ways to sustainably invest in natural capital such as forests, fisheries, water, and agriculture; how to improve food security, improve nutrition, and reduce poverty; efficient ways to invest in resources and energy efficiency; and to overcome obstacles and embrace opportunities associated with renewable energy. And finally, the report examines: how to make the *shift* to a green economy; how financing works, how it is modeled, and what conditions are necessary for a success.

Getting Involved

Although the government has passed laws to protect the environment, private citizens have assumed a sense of personal responsibility and taken measures to live sustainably and reduce waste on their own.

A noteworthy trend is to use the natural landscape and native vegetation, as opposed to the water-thirsty English gardens that homeowners once preferred. Given water demands and arid climates, homeowners and cities have started growing native gardens with only vegetation native to that given area, minimizing the use of water that would otherwise get to lawns and water-thirsty plants.

Many homes and office buildings are now being built to maximize use of natural energy and sunlight. This is known as **green architecture**. Norman Foster designed the Gherkin building in London—a prime example for green architecture. This structural masterpiece was specifically designed with a curved exterior to maximize its own ventilation and to become less reliant on air conditioning. Its glass exterior and open floor plans maximize ambient sunlight, so it is less reliant on electrical lighting.

Katherine Welles/Shutterstock.com

A LEED Certified Building

A popular trend of the past 25 years has been to build homes as large as possible. But a new trend is taking hold and that is to build smaller homes for quality instead of quantity. Interior designers and architects seek to capitalize on this trend, and books are published on how to maximize space and use renewable energy to meet the home's electrical needs. Reclaimed lumber and recycled copper are often used.

LEED (Leadership in Energy and Environmental Design) **certification** is a voluntary program whose participants (buildings and home owners) get certified by an objective third party as being considered green. This program has now spread to 135 countries.

LEED homes and buildings have become a desired commodity because they are designed to reduce waste, use local resources, conserve water and energy, lower greenhouse gas emission, and provide an overall healthy environment to its occupants. Highly sustainable buildings earn a platinum status and now there are even 'living buildings' that are carbon neutral!

Chevron Corporation's building in Louisiana is LEED certified. The College of William and Mary has constructed LEED certified buildings as well as many new buildings in the University of California (UC) system. The UC system has estimated that it has already saved millions of dollars with its sustainable practice.

tlorna/Shutterstock.com

Green Architecture

The World Takes Action

In 2002, an UN-sponsored coalition of businesses, foundations, heads of state, and dignitaries decided to start a campaign and take action to achieve eight Millennium Development Goals (MDGs) originally by 2015, and now beyond. Various fundraisers, committees, and summits have been held to spearhead the cause and reach its set goals. The eight MDGs are listed below:

Hung Chung Chih/Shutterstock.com

1. Eradicate extreme hunger and poverty
2. Achieve universal primary education
3. Promote gender equity
4. Reduce infant mortality

5. Improve maternal health
6. Combat HIV/AIDS and other diseases
7. Ensure environmental sustainability
8. Establish global partnership

Every year, an MDG report is generated to provide yearly assessments of how much progress the globe has achieved and what strides it is taking to reach its MDGs. Although there is still a lot of focus, determination, and hard work ahead, the following are encouraging statistics surrounding the progress of each of the Millennium Development Goals:[iii]

Sustainable Development Goals

Because of the great success of the UN's Millennium Development Goals for 2000, the United Nations spearheaded a new initiative that built on the success of the MDG's; a global rally cry to protect the environment, end poverty, and ensure a quality life for all of the planet's inhabitants and adding the planet as a priority. From this initiative came the 'Sustainable Development Goals' or SDGs.

The SDGs are as Follows:

SDG1. End poverty
SDG2. Zero Hunger; fight food insecurity, promote sustainable agriculture and healthy nutrition
SDG3. Health and Well-being for everybody
SDG4. Inclusive and equitable education
SDG5. Gender equality and the empowerment of women
SDG6. Clean water for all, and sustainable water management
SDG7. Affordable, renewable, clean energy is available to everybody
SDG8. Sustainable work growth and employment
SDG9. Build resilient infrastructure and sustainable and inclusive industrialization
SDG10. Reduce inequality within and among countries
SDG11. Sustainable cities and communities
SDG12. Promote responsible consumption and production
SDG13. Combat climate change and take climate action
SDG 14. The conservation and sustainable use of the oceans and all marine ecosystems
SDG 15. Promote sustainable terrestrial ecosystems by halting land degradation, biodiversity loss, sustainable use of terrestrial ecosystems and sustainable forest management
SDG16. Promote peaceful societies with sustainable development
SDG17. Strengthen global partnership for sustainable growth
(source www.undp.org)

Review Questions

1. Why are survivorship curves important for extrapolating information about a given population? Characterize Type I, II, and II survivorship curves and provide examples.

2. Summarize the difference between R and K-Selected Species. Draw growth curves for each. What is carrying capacity? What is biotic potential?

3. Define TFR and "Replacement Rate." What factors may account for the worldwide TFR being reduced?

4. Describe the difference between classical economics, environmental economics, and ecological economics.

5. What is the significance of OSHA? NEPA?

6. What is UNEP's view of the economy?

Discussion Questions for Class

A. How have the root causes of environmental problems influenced population growth and population demographics around the world?

B. Population demographics: as baby boomers age, there may be more dependents than independents. How does this affect the work force and politics?

C. How does overpopulation affect health care and overall quality of life?

D. Since China has implemented the one child policy, women have become a stronger presence in the work force, which has, in turn, improved their quality of life. How does education affect quality of life? How does poverty affect birth rates in women?

E. Have we reached carrying capacity? How did you arrive at this conclusion? What factors influenced the biotic potential on our growth curves? How about carrying capacity?

F. Critics of microlending argue that it done little to promote gender equality and ineffective in breaking the cycle of poverty. What are your thoughts on this?

G. How might unplanned urbanization lead to significant environmental problems?

H. How have the root causes of environmental problems influenced local and international politics?

I. Should there be an international agency overseeing environmental problems?

J. Discuss pros and cons of such an agency.

K. How have the root causes of environmental problems increased our need for green architecture and landscape?

L. Are we headed toward a global green economy? Why or why not?

M. How green is your home? What improvements can be made?

N. How green is your landscaping? What improvements can be made?

O. Getting approval for EIRs can be time consuming and even stall major projects. Given that many jobs are at stake, how strict should public agencies be about allowing projects to progress?

P. Can you think of different ways in which worker co-ops are environmentally sustainable?

Your Personal Sustainability Journal

Set up your utility bills and all accounts so that you pay them online. Go wireless!

Moving creates a lot of waste! If you plan to relocate in the next 6 months, cut down on all shopping such as buying food and toiletries. That way, there will be less to discard when you move. Look into companies that recycle moving boxes.

Donate to a literacy charity or donate to help raise schools or donate textbooks to a library.

Volunteer to tutor young children. Volunteer at the local library and volunteer your time to read stories to children. (And great for the resume!)

Set academic goals for yourself and follow through!

Refuse condiments or flatware that you will not need when ordering takeout.

Turn off lights when they are not being used.

Add your own suggestions here.

Revisit some of the websites that were mentioned in the first chapter: www.nrdc.org, www.ran.org, and www.ucsusa.org. Strengthen your resolve and get involved in their causes.

Plant or rent a plot at a community garden!

Get involved! Volunteer. Read to children. Tutor. Clean up beaches and parks.

Educate your family, friends, and coworkers on the environment and sustainable living.

Support the “Green” party.

Spend more time outside during the summer.

Write a letter to your local congressman or newspaper expressing concern and asking for more actionable proposals.

Write letters commending businesses for their conservation efforts. Share this on social networking sites.

Recycle as often as you can! Newspapers, cans, plastic, and batteries. Make it a lifestyle commitment!

Use heavy drapery to keep heat in during the winter and nighttime, and to keep the sun out when it is hot.

Be the kind of responsible, environmentally conscious person that you would like the rest of the world to be!!!

Do not underestimate the power of grass-roots effort.

Add your own suggestions here.

Endnotes

i. From the CIA World Factbook www.cia.gov
ii. From The CIA World Factbook www.cia.gov
iii. http://www.nobelprize.org
iv. For additional reading, see Banker to the Poor: Micro-Lending and the Battle Against World Poverty by Muhammed Yunus
v. For more information, read "An Introduction to Ecological Economics" by Robert Costanza, John Cumberland, Herman Daly, Robert Goodland, and Richard Norgaard
vi. www.unep.org
vii. http://www.un.org/millenniumgoals/

Glossary

abiotic Describes phenomena such as sunlight, precipitation, and wind that are not related to or derived from life or living organisms. Contrast to **biotic**.

abyssal zone The oceanic zone found 2,000–4,000 meters down with cooler temperatures (less than 5°C) and very little light. One finds luminescent fish, squid, and eels in the abyssal zone.

acid A substance with a pH lower than 7, thereby consisting of a high concentration of ions. Acids are corrosive and caustic, have a sour taste, and turn blue litmus to red.

acid mine drainage The sulfuric acid produced by **surface mining** that can run into water supplies.

acid rain A by-product of burning fossil fuel, it is rainwater that has a pH of 5.7 or less. It is damaging to aquatic life, vegetation, buildings, statues, and other infrastructures.

acidic Said of aqueous solutions with a pH balance below 7 and giving off an excess of H+ ions.

acidification Phenomenon occurring as a result of carbon dioxide being absorbed by our oceans, creating carbolic acid in vast quantities, thus lowering their pH to an acidic level.

active solar energy A process that uses conductors such as metal to trap sunlight.

adaptations Structures of behaviors that increase the livelihood or **fitness** of an individual. Over time, structures, and genetic frequencies shift so profoundly that a new species forms.

adaptive immunity The method that the immune system develops to defend the body against infection. It is based on its reaction to pathogens it has been exposed to.

adaptive radiation Species occupying new niches and adapting uniquely to each niche, ultimately causing a great deal of biodiversity.

adenosine triphosphate (ATP) The molecule that is the energy currency of a cell.

adhesive Characteristic of polar molecules (such as water) that compel them to bond with both positively and negatively charged particles as well as like particles.

affluence Generally, having and using means in excess of basic needs. It is a relative term, as even the poorest westerner may appear to have considerable wealth compared to the average person of the developing world. In the context in which we are using the term, affluence leads to excess consumption and excess waste. It means a lot of air travel on planes, private jets, and large luxury vehicles. These all use a great deal of fuel and leave a large carbon footprint. Wealth leads to an overconsumption of gadgets and electronics that leave excess toxic, non-degradable waste and increased pollution. Wealth also means overdevelopment of land for large homes and shops.

agroforestry A diverse, highly functional, and integrative approach to agriculture that combines crops with shrubs, trees, and livestock.

The Alaska National Interest Lands Conservation Act A 1980 act that designates large parts of Alaska as national forests, parks, and wildlife refuges.

Allee effect The theory that a low density of population of an organism can have a negative impact on net population growth.

alley cropping (same as strip farming) The practice of growing crops in rows to keep pests confused and to maintain soil fertility. These rows are also typically flanked by tall vegetation in order to prevent erosion caused by excess wind.

allopatric speciation Speciation that occurs in the presence of a physical barrier or geographic isolation.

ambient-light hypothesis Hypothesis that suggests the more sunlight accessible to an ecosystem, the more functional it becomes.

amensalism A type of symbiosis in which one species is harmed and the other is unaffected.

amino acids The building blocks of proteins.

analogous structures Anatomical structures that do not share an evolutionary past but are the result of similar environmental selective pressures.

anion A negatively charged ion.

aposematic coloration A characteristic in which the coloration of a species is beneficial to both the species and a potential predator by warning away the predator.

aposematic coloring Bright coloring in a species that warns potential predators of toxicity and other defenses.

anaerobic respiration Creating energy (splitting sugar) in the absence of oxygen.

anchorage dependence Condition of healthy cells that stay anchored to their place of origin and do not invade other tissue.

angiogenesis The creation of vascular tissue; in cancer cells, it is used to feed tumors.

angiosperms Flowering plants and trees.

apoptosis Cell suicide necessary in the natural life cycle of an organism.

aquaculture Fish farming.

aquatic ecosystem An ecosystem that exists in water rather than on land.

aquifer Underground reservoir of water that has seeped into limestone rock cutouts.

asthenosphere The warm less dense layer found below plates.

atmosphere A compilation of gases that surrounds the earth.

atom Smallest unit of matter that retains its own chemical properties.

atomic mass The mass of protons plus the mass of neutrons in an atom.

atomic number The number of protons in the nucleus of an atom.

ATP (adenosine triphosphate) The molecule that is the energy currency of a cell.

autotrophs Organisms that produce their own energy and weigh in at the first **tropic level**. Synonym: producer.

background extinction Extinction of a relatively long-standing species caused by moderate changes in the environment.

basic Said of an aqueous solution with a pH balance over 7 and giving off an excess of OH^- ions.

batesian mimicry Coloration or other characteristics of a species that indicate to potential predators that it is harmful or unpalatable when, in fact, it is not. Contrast with aposematic coloring.

bathyal zone The oceanic zone that contains cooler temperatures (roughly 5–10°C) and decreasingly less sunlight. The type of marine life found here is classified as nekton—strong swimmers such as sharks and whales.

benthic zone Biodiversity rich region at the bottom of a body of water.

bioaccumulate Said of pesticides and chemicals that do not degrade over time and gather in the food chain.

biochemical record In the Chapter 3 discussion, this information indicates that all living organisms share the same molecules that form genetic material, from bacteria to cows.

biodiversity The many varied species of a given area.

biogeography The study of living things throughout the planet.

biological extinction When a species is no longer found anywhere on earth.

biomagnification Bioaccumulation of pesticides and chemicals that resist degradation in large predatory fish.

biomes Observable trends of vegetation in terrestrial ecosystems that are directly correlated to climate.

bio-mining Mining that utilizes microbes to break down **overburden** in order to reach the minerals beneath.

bioremediation A form of green technology that uses microorganisms such as bacteria and protozoa to break down toxic chemicals in the soil and biotransform them into CO_2.

biosphere The entire surface of the earth with consideration of its **biotic** and **abiotic** factors; the living components of the earth in their entirety.

biotechnology The science of manipulating living organisms for human need and benefit.

biotic Concerning or produced by living organisms. Contrast with **abiotic**.

biotic potential The possible growth rate of a population of organisms under ideal conditions.

bitumen A type of semisolid crude oil containing excessive impurities.

boreal forest (taiga) Biomes of vast spruce, fir, and pine forests that exist in harsh and cold climates in the northern hemisphere.

bubo Swelling of the lymph nodes.

buffer A substance that helps maintain a consistent pH. Blood acts as a buffer in our bodies.

bushmeat Meat from wild animals.

bycatch Unintended catch in trawlers' nets.

cap and trade A practice in which companies receive a cap or limit on the amount of carbon dioxide or other pollution they can release into the atmosphere. As a company's pollution decreases beneath its limit, it can sell **pollution credits** to other companies that exceed their limits.

carbohydrates Sugars. They include Glycogen in animals and starch in plants.

carbon footprint The amount of carbon (in the form of methane or carbon dioxide) released by a given **population**.

carbon monoxide A colorless and odorless gas that is released from car exhaust, and forest fires and household heat producers such as heaters, furnaces, and ranges.

carbon sink Any storehouse for carbon.

carbonic acid Acid produced when the carbon from the burning of coal reacts with water vapor in the atmosphere.

carcinogen Any cancer-causing agent.

carcinoma A form of cancer that originates in epithelial tissue.

carrying capacity The maximum population of a species that an environment can accommodate; beyond this point, overpopulation becomes problematic.

Cartagena protocol An international treaty ensuring the safe handling, care, and transport of organisms derived from bioengineering and biotechnology.

cation A positively charged ion.

cation exchange The process in which plants obtain nutrients.

cell The basic unit of life.

cellulose A carbohydrate that makes up plant cell walls. We know it as dietary fiber.

cell theory Theory that states that (1) living things are made up of cells; (2) cells are the most basic unit of any organism; and (3) existing cells are derived from pre-existing cells.

CEQ **Council of Environmental Quality,** established to oversee the efforts of NEPA.

chains of extinction The connection of species as they relate to one another as predator and prey.

chaparral Biomes that consist of Mediterranean-type climates and are characterized by long, hot summers and cool, wet winters. Also known as **Mediterranean woodlands**.

chemical Used to describe components of the environment that include such factors as the molecules that make up the atmosphere or the composition of soil or water in a given ecosystem; the chemical components of the environment are on a chemical (microscopic) level.

chemical energy Energy stored in the bonds of atoms.

chemosynthesis A process in which organisms obtain nourishment from chemicals released from ocean vents.

chemotherapy Type of cancer treatment involving chemicals that target dividing cells.

chlorophyll A green pigment found in photosynthetic organisms that can convert sunlight to energy.

chloroplasts Subcellular structures found in plants where photosynthesis takes place.

CHNOPS An acronym for; Carbon, Hydrogen, Nitrogen, Oxygen, Phosphorous, and Sulfur, the most abundant elements found in living things.

cilia Small hair-like projections in respiratory systems whose function is to sweep out irritants and impurities, thus protecting our lungs.

Clean Air Act A 1963 US Federal law designed to control the country's air pollution. It was strengthened by amendments in 1967, 1970, 1977, and 1990.

clear cutting The practice of cutting down all the forestation in a given area in one fell swoop.

climate Long-term trends in weather patterns of a large given land area.

climate change A potentially catastrophic change in weather patterns and temperature trends worldwide. Climate change occurs as a result of human activity as well as of natural phenomena.

climax community The stable ecosystem that evolves to become rich in flowering plants, trees, biomass, and biodiversity in general.

co-evolution A phenomenon in which two species put selective pressure on each other and influence the other's evolution. An example would be the relationship between a plant and its pollinator.

co-generation Combined heat and power for the generation of electricity and heat. Rewrite this one.

cohesive Having the electromagnetic attractive characteristic of polar molecules which bond as a result of having both positive and negative charges.

cohorts Individuals of a given population born at the same time.

combustible items Items and chemicals with a strong capability of igniting or burning at normal temperatures.

commensalism Symbiosis in which one species benefits while the other is neither harmed nor benefited.

communicable disease A disease transmittable between hosts.

community Multiple populations interacting in a given area.

competition An important form of community interaction that includes species and individuals vying for space, mates, food, and other resources.

competitive exclusion A scenario in which the species that is the superior competitor will drive out or kill off the less fit species.

compounds Molecules that are made up of different elements always found in a consistent ratio.

concentrated solar power (CSP) a form of active solar energy that uses hundreds and sometimes thousands of mirrors to concentrate sunlight and amplify its reflective properties.

coniferous forest Biomes of cone-bearing trees that typically flourish in the northern hemisphere's harsh and cold climes, but can also thrive in temperate zones.

conservation The act of preserving or protecting an ecosystem, its species, and its biodiversity.

consumers Organisms that must consume other organisms for energy. Synonym: **heterotroph**.

contact inhibition Characteristic of healthy cells that, when dividing and come into contact with too many cells, they stop dividing.

contour farming The practice of planting crops to go along with the gentle rolling slopes of hills instead of orienting them perpendicular to inclines, thereby reducing soil erosion.

contour mining A type of surface mining that extracts coal found in exposed seams on mountainsides. This type of mine is smaller than most but generates considerable waste.

control Replication of an experiment without the variable that is present in the original experiment.

convergent evolution Evolution that causes unrelated species to look or behave similarly as a result of similar environmental adaptations. It is a result of these species occupying similar environments so that natural selection favors similar adaptations.

convergent boundary Collision between tectonic plates. When a **convergent boundary** occurs beneath the ocean, the thicker oceanic plate slides beneath a thinner continental plate.

coral bleaching The phenomenon when corals cease to have a symbiotic relationship with algae, typically due to stress. As a result, they lose their color.

coral reefs Reefs found in shallow, temperate marine habitats that receive plenty of sunlight. Here, sea anemone-type creatures called polyps secrete calcium bicarbonate that creates the corals on the shallow ocean floor, home too many species of algae and other marine life. The corals forge a mutualistic, symbiotic relationship with the algae to create the beautiful hues of the coral reefs.

Coriolis effect The tendency of prevailing winds in the northern hemisphere to deflect to the right, while prevailing winds in the southern hemisphere to deflect left. This tendency is a product of the rotation and curved surface of the earth.

corrosive items Reactive chemicals that damage or compromise the surface of a material.

Counsel on Environmental Quality (CEQ) An arm of the US executive branch whose duty is to coordinate efforts to protect and maintain overall environmental quality.

covalent bond Atoms in close proximity that share an electron and combine to become part of a larger molecule.

crop rotation A long-range system of growing dissimilar crops in a limited area, changing the location of each crop from season to season. This practice maintains soil fertility and enables crops to evade pests naturally.

crown fire A forest fire that burns hot and hops from crown to crown of elevated terrains, killing trees and wildlife and resulting in soil erosion.

CSP **concentrated solar power,** a form of active solar energy that uses hundreds and sometimes thousands of mirrors to concentrate sunlight and amplify its reflective properties.

dead zones Areas with low biodiversity.

deciduous forest Biomes that are comprised of hickory, maple, and oak trees that lose their leaves annually and thrive in colder temperate zones.

decommissioned Necessary process of maintaining nuclear power plants after they become no longer productive.

decomposers Such as fungi and earthworms, are the recyclers of an ecosystem. They break dead organic matter into simple atoms and release them back into the environment.

degenerate stage The final stage of the cycle of change of an eco-system when much of the biomass starts to die and decompose. This stage produces a great deal of dead, organic matter. Lichens and bryophytes (moss) are common, replacing the more mature plant life.

denitrification Process in which decomposers break-down dead organic matter, and nitrogen gas is released into the atmosphere.

deoxyribonucleic acid (DNA) Molecule that stores the genetic blueprint of a cell or organism.

desalinization plants A facility whose purpose is to remove salt from ocean water.

desert Biomes where evaporation exceeds precipitation.

desertification Any human activity that strips soil of nutrients (such as deforestation, excessive grazing, or ill-managed cultivation) that causes desert-like conditions. Desertification makes it difficult for plants and wildlife to grow and flourish, as soil can no longer hold nutrients.

desertified Said of the condition of semi-arid soil which can no longer hold nutrients.

deterministic extinction The assumption that a species will go biologically extinct unless changes, such as preserving habitats and conducting aggressive conservation efforts, are implemented.

detritus A layer of dead, organic matter covering the ground.

detrivores Such as beetles and maggots, feed on dead organic matter.

developed nations Industrialized nations with stable economies.

developing nations Nations with unstable economies that are in the process of industrialization.

disturbance Any event that compromises or causes significant change to an intact ecosystem.

divergent boundary A fault that occurs beneath the ocean as a result of two plates moving apart at the divergent plate boundary, lifting the two plates.

DNA (deoxyribonucleic acid) Molecule that stores the genetic blueprint of a cell or organism.

EA's (Environmental Assessments) Assessments carried out by the **CEQ** that examine whether construction or human activity would have potential negative impacts on the environment and overall quality of life.

echolocation The navigation method used by bats and dolphins that involves emitting a sound and using the echo to analyze the environment.

ecological extinction Extinction of a species' population that occurs when its size becomes so small, it cannot perform its ecological role in an ecosystem.

ecological footprint The amount of land and resources required to supply a population with its goods and services and absorb and assimilate its waste.

ecology The study of the living world and its interactions with the **abiotic** environment.

ecosystem A community, with consideration of its **biotic** and **abiotic** factors.

ecosystem diversity The variety within an ecosystem that encompasses different biological communities, ecological processes, and various habitats.

ecotones Areas where two distinct Biomes overlap.

edge effect Desertification of land from excessive sunlight caused by clear cutting forests.

EIS (Environmental Impact Statement) An analysis, prepared after an EA is found positive, which examines all possible issues related to the project—particularly, how it will impact both the environment and quality of human life.

electromagnetic energy Energy in the form of electrical and magnetic waves traveling through space.

electron Negatively charged particles found circulating around the nucleus of the atom.

element The smallest unit of matter (made up of atoms) that retains its own unique properties and cannot be broken down by normal physical means, such as heating.

emerging virus A newly discovered virus that is increasing, or has the potential to increase, in frequency of incidence.

endangered Status indicating that a species is about to go extinct.

endangered species Species with such low abundance that they are at high risk of becoming extinct.

Endangered Species Act of 1973 A US law that established that it is illegal to harm or kill an endangered species.

endemic species Species native to a given area.

endergonic reaction A metabolic reaction that has a net positive change in free energy.

energy The capacity to do work. Energy is either **potential** or **kinetic**.

environmental assessments (EAs) Assessments carried out by the **CEQ** that examine whether construction or human activity would have potential negative impacts on the environment and overall quality of life.

environmental economics A subdivision of economics, applying the environment to its economic model. Environmental economics supports the notion that many of our natural resources come from the environment, and we can use them to build an economy with its capital, but we must acknowledge its limitations and use it sustainably.

environmental impact statement (EIS) An analysis, prepared after an EA is found positive, which examines all possible issues related to the project—particularly, how it will impact both the environment and quality of human life.

environmental science A multidisciplined study of human interaction with the living and the nonliving world that encompasses sociology, physics, ecology, geography, humanities, economics, political science, etc. Its goal is to examine and explore the many environmental issues that we face today.

environmentalism A social movement committed to conserving or improving the environment and to living sustainably.

enzyme A biological catalyst.

epiphytes Plants that live on the tops of trees (especially atop the canopies of rainforests). They also grow on plants and on stationary objects such as buildings.

erosion The mobility of soil, caused by wind and water, which ultimately strips it of its nutrient content.

estuaries Areas where freshwater meets saltwater, or, more specifically, where rivers empty into the ocean.

ethology The scientific study of animal behavior.

Eukaryotic cells Associated with multicellular organisms. These cells contain a nucleus and membrane-bound organelles.

euphotic zone The oceanic zone that sits atop a continental shelf and receives the most net sunlight. Many photosynthetic organisms such as algae and phytoplankton are found here, as well as nonphotosynthetic zooplankton. In the tropical climates, the euphotic zone contains the coral reefs, home to sea creatures such as anemones and sponges.

eutrophic Describes aquatic environments containing little Oxygen concentration.

eutrophic lakes Lakes with low supplies of oxygen.

eutrophication Having little oxygen supply in lakes. Often caused by man-made pollution.

evolution The change in gene frequency of a given population over time. It is important to note that populations evolve, not the individuals within a population.

ex situ conservation 1. The removal of organisms from natural habitats to human controlled areas such as zoos, animal parks, seed banks, and adaptive breeding situations; or 2. The process of protecting organisms by transplanting them from a threatened environment to an alien but more viable natural environment.

exergonic reaction A metabolic reaction that releases heat.

exploitation Community interaction in which one species exploits another for its own benefit. Parasitism, predation, and herbivory are forms of exploitation.

extirpated Condition of a species which has succumbed to a local extinction.

extraction the human activity of taking goods and resources from the earth for consumption.

fault An area where two masses of the earth's mantle rub together as a result of the earth's internal forces.

fauna All of the native animal life of an area.

fermentation A type of anaerobic respiration that releases ethanol as a by-product.

first law of thermodynamics Energy cannot be created or destroyed, it can only change form.

first trophic level Beginning of the food chain. Plants and algae which are self-sustaining (in that they exist through photosynthesis) make up this level. Other words for these organisms are **producers and autotrophs.**

the Fishery Conservation and Management Act A 1975 act passed to protect the U.S. coast from foreign fishing up to 200 miles out from the coastline.

fitness Structures of behaviors and characteristics that improve the adaptive qualities of an individual.

flagella A tail-like structure characteristic of bacteria and sperm cells made of protein and used for mobility.

flagship species A species, such as the polar bear, that has become a symbol for a conservation movement.

flora Naturally occurring plant life of an area; native vegetation.

food insecurity Condition of not getting enough to eat to maintain health.

fossil record Evidence derived from fossils found in strata of rock that provides evidence of when, where, and how organisms have lived in the earth's history.

foundation species Species that create habitats for other species. Coral and beavers are examples.

fracking A method of harvesting natural gas that involves using water, sand, and chemicals to break up rock, build up pressure, and release natural gas from rock beds.

freshwater habitats Habitats such as streams and lakes that do not have high levels of salinity.

fundamental niche The full potential range of all factors that a given species would encounter in its biotic environment during the course of its lifetime. These factors include, but are not limited to, the following: altitude, pH, diet, and thermoregulation.

gene banks Storage facilities where seeds are stored in low temperatures and low humidity. Also known as **seed banks**.

generalist species Species that occupy a wide niche range. They adapt readily to environmental changes.

genetic diversity The variety of genes within a species. Genetic diversity is beneficial to a species because it helps the species adapt to a changing environment.

genetic engineering The process of altering the DNA of a living organism to create products of human interest.

genetically modified organisms (GMOs) Genetically engineered organisms.

Geophagy The practice of eating soil, most often clay, in order to gain nutrients, especially minerals, to address a diet deficiency.

geothermal heat pump Device that utilizes fluid-filled pipes that are run underground by engineers that reach the same temperature as the ground. The fluid is pumped into homes and transferred into the home's air ducts and circulated.

glaciers Large bodies of ice which result from the ice melt of warm seasons not keeping up with the increase of ice during cold periods. Glaciers can extend, in size, from hundreds of square yards to thousands of square miles, and, in duration, over decades, centuries, or even millennia!

global distillation Also known as the **grasshopper effect**, an atmospheric phenomenon by which pesticides, toxic chemicals and heavy metals manufactured in warm climates heat and rise in gaseous states and are carried to colder climates.

global warming An overall increase in atmospheric and oceanic temperatures due to the greenhouse effect.

GMOs, (genetically modified organisms) Genetically engineered organisms.

grasshopper effect Also known as **global distillation**, an atmospheric phenomenon in which pesticides, toxic chemicals, and heavy metals manufactured in warm climates heat and rise in gaseous states and are carried to colder climates.

grasslands Biomes that are found in the interior of continents where the annual rainfall is too little to sustain forest growth, but too abundant for a desert biome.

green architecture Architecture built to maximize use of natural energy and sunlight.

green infrastructure Mimicking natural processes, this means of addressing water pollution uses vegetation and soil to capture and enhance the hydrologic cycle to prevent excess runoff in urban areas.

green technology A means that uses natural organisms to address pollution.

greenhouse effect The phenomenon created when atmospheric gases trap infrared radiation as heat, reradiate it back to earth, and contribute to the overall warming of earth's surface.

greenhouse gas Any atmospheric gas that is linked to global warming.

groundwater Water that has percolated into the ground.

gyres Large systems of rotating ocean currents. They involve wind and the Coriolis Effect.

HAA (hormonally active agents) chemicals that often mimic estrogen, interfering with the endocrine systems of mammals.

habitat destruction Unsustainable human practices that take habitats away from plants and animals.

habitat fragmentation A dramatic reduction in the size of an area of a species' habitat.

habitat loss Phenomenon in which some native species of an area are displaced by the needs of an expanding human population.

hadal zone The deepest oceanic regions found in ocean trenches. Hadal zones are rich with organisms that feed on chemicals released from ocean vents.

Heterotrophs Organisms that must consume other organisms for energy. Synonym: **consumer**.

heterozygous Possessing two different genes for a specific trait.

high specific heat The characteristic of a substance that requires a great deal of heat to raise the temperature of the substance to rise 1°C. Water has a high specific heat.

HIPPCO Acronym used as a mnemonic to remember all the threats to biodiversity: H for Habitat loss; I for invasive species; P for Population (as in overpopulation); P for Pollution; C for Climate change; and O for Overexploitation.

homeostasis A consistent internal environment that allows life to continue without need for drastic adaptation.

homologous structures Anatomical structures common to various species that provide evidence of a shared evolutionary past.

hormonally active agents (HAA) chemicals that often mimic estrogen, interfering with the endocrine systems of mammals.

homozygous Possessing two of the same genes for a specific trait.

hot spots Areas of high biodiversity.

hydroelectricity Electricity produced in dams that employ turbines to convert steam.

hydrogen bonds Bonds or attraction between water molecules (H_2O).

hypothesis Generally, a statement or prediction that is to be verified or disproved through data collection and observation.

ice age Long periods of cooling on the earth when ice sheets are widespread and vast, and when extended glaciation occurs.

igneous rock Rock that is cooled magma, usually formed through volcanic activity.

indicator species A species whose level of well-being is indicative of the health of its ecosystem.

industrial farming The use of technology in agriculture.

industrialized nations Countries in which the economy is driven mainly by manufacturing and trade, as opposed to agriculture.

infectious disease A disease that is transmittable between hosts.

infrared Long range light (long wavelengths) emitted from the sun.

inorganic molecule Molecules that compose chemicals that are not carbon based. They can occur naturally but are often synthetic. Inorganic chemicals are potentially harmful to the body because the body does not recognize them and cannot break them down properly. Many industrial chemicals, fertilizers, and pesticides are inorganic and contain many potentially harmful side effects to our health and the health of the environment.

in situ Latin for "in place."

in situ conservation Efforts to maintain and promote the survival of organisms (plants or animals) in their natural habitat.

integrative pest management (IPM) approach to agriculture that maximizes pest control by employing many various integrative methods that effectively minimize the use of pesticides, and adjusting the use of each method accordingly.

Intergovernmental Panel on Climate Change (IPCC) a Nobel Peace Prize-winning panel overseen by the UN whose priority is to examine and assess climate change and its impacts through a scientific lens. Scientists participate on a volunteer basis.

intersexual selection A characteristic behavior in which one sex of a species, typically the female, is discriminate in looking for suitable traits in the opposite sex.

intertidal zones Dynamic shallow-water marine habitats characterized by sandy beaches.

intrasexual selection A characteristic behavior in which members of the same sex of a species compete with one another.

introduced species Also known as exotic species, these are species that are introduced into a given area from other parts of the world.

invasive species Invasive (sometimes called exotic) species are species from other parts of the world that are introduced to a new habitat.

ionic bond A bond created by the attraction of a negatively charged anion to a positively charged cation.

ions Atoms or molecules with either a positive or negative charge.

IPCC (Intergovernmental Panel on Climate Change) A Nobel Peace Prize-winning panel overseen by the UN whose priority is to examine and assess climate change and its impacts through a scientific lens. Scientists participate on a volunteer basis.

IPM (integrative pest management) approach to agriculture that maximizes pest control by employing many various integrative methods that effectively minimize the use of pesticides, and adjusting the use of each method accordingly.

isotopes Molecules whose atoms have varying numbers of neutrons. Isotopes are often unstable and decay, sometimes becoming radioactive.

iterative process Activity, experimentation, and investigations that are varied and repetitious, and are carried out with a desired goal.

keystone species A species whose presence and well-being hold great significance in maintaining equilibrium of the population of all species in its community.

kin selection The phenomenon that the probability of altruistic behavior within a species increases directly according to how closely related a population is.

kinetic energy Energy in motion. Once the diver jumps and activates the spring, the energy in that spring has changed from potential to kinetic energy.

k-selected species Species that have longer life cycles, fewer offspring, and mothers who put more care into raising their young. Elephants, humans, and other primates are examples of K selected species.

Kyoto Protocol The agreement by multiple nations in Kyoto, Japan in 1997 which negotiated to address escalating concerns of climate change. Its goal was to discuss, devise, and enforce measures for industrialized nations to monitor greenhouse gases and carbon emissions. The goal negotiated was to reduce these emissions by 5% (relative to 1990's standards) which, due to increased carbon dioxide concentrations, translates to almost 30% reduction by present day standards.

lakes Basins or depressions in land that collect and act as a storehouse of water from runoff, mountain glaciers, lakes or streams. Lakes are usually fresh water but may have high concentrations of salt, as does the Great Salt Lake.

lead A heavy metal found both naturally and in manufactured goods, but most concentrations in the air are from automobile fuel.

Leadership in Energy and Environmental Design (LEED) a voluntary program whose participants (building and home owners) get certified by an objective third party as being considered 'green.'

LEED certification (Leadership in Energy and Environmental Design) a voluntary program whose participants (building and home owners) get certified by an objective third party as being considered 'green.'

leukemia A form of cancer that originates in bone marrow.

life table Graph that indicates age of death. Categories include Type I (death rate highest in the elderly), Type II (death rate is consistent throughout the populations' life cycle), and Type III; (death rate is highest among juveniles).

limiting factors Factors that prevent an organism from surpassing its **realized niche** and from realizing its **fundamental niche**.

limnetic zone is the upper layer of open water in a lake or pond.

lipids Lipids are fats, hydrophobic biological molecules that make up cell membranes and store energy.

lithosols Saline soils.

lithosphere The outer, rocky, and solid surface of the earth.

littoral zone is near the edge of a lake or pond.

local extinction Extinction that occurs in one part of the world but not elsewhere. The Black Rhinoceros is an example.

macronutrients Nutrients the body needs in large amounts such as protein, carbohydrates, and essential fats.

magma The liquid and semiliquid state that the earth's rock **(mantle)** acquires being exposed to temperatures up to 9,000°F at the earth's core.

mantle The 29,000 km thick layer of earth (its thickest) made up of semisolid rock located beneath the earth's crust and above the earth's core.

marine habitats Habitats such as oceans that have high levels of salinity.

Marine Stewardship Counsel (MSC) A nonprofit, international agency that works with scientists and fisheries to regulate and set standards to ensure fishing practices are carried out sustainably.

mass extinction Extinction of a vast number of species in a short period due to rapid climate change, over-exploitation, deforestation, or general loss of habitat for many species.

matter Anything composed of atoms and molecules; anything that has mass and takes up space.

mediterranean woodland Biomes that consist of Mediterranean-type climates and are characterized by long hot summers and cool wet winters. Also known as **chaparral**.

mesosphere The third atmospheric layer which extends some 30–50 miles up from the earth's surface and sustains temperatures as low as 290°C!

metamorphic rock Rock such as marble that changes form due to heat, pressure, and time.

metastasis Condition in which cancer cells enter blood or lymph, becoming systemic.

micro-lending A banking system that involves lending money to the poverty stricken with no collateral in hopes that borrowers will become financially independent, breaking the cycle of poverty.

micronutrients nutrients the body needs in small amounts such as iodine, iron, and vitamins.

microbes A microscopic organism.

microfiltration A process in which water is pushed through a membrane-like screen which mechanically separates water from salt.

minimum viable population (MVP) The minimum number of individuals in a population of species in the wild that will provide enough genetic variability to sustain its population.

mitochondria A subcellular structure that acts as the powerhouse of the cell by generating ATP.

molecular formula Combination of symbols that represent the elements comprising a compound and the number of each element's atoms required to compose a molecule of that compound.

molecule A group of atoms that are bonded together to form a larger group of molecules. Collectively, the larger group of molecules has its own unique properties that are different from those atoms of which it is comprised.

monoculture A practice in which one crop is grown at a time in a given area.

montreal Protocol An agreement reached in 1987 by an international panel in Montreal, Canada, designed to cut down on chlorofluorocarbons (CFCs). This protocol has been carried out with considerable success.

MSC (Marine Stewardship Counsel) a nonprofit, international agency that works with scientists and fisheries to regulate and set standards to ensure fishing practices are carried out sustainably.

mucus A thick, sticky substance found in the nose and other body cavities whose job is to trap and flush out dust and other impurities.

mullerian mimicry A similarity of characteristics in different species that warn predators that these species are harmful or unpalatable.

mutualistic symbiosis When all parties benefit from living together.

MVP (minimal viable population) is the minimum number of individuals that must be present in a species population for it to resist extinction.

Nagoya Protocol An international treaty ensuring the "fair and equitable sharing" of benefits arising from genetic resources.

nanotechnology A method of addressing water pollution that involves manipulating the environment on a molecular level.

National Environmental Policy Act of 1969 (NEPA) A U.S. law designed to protect and maintain overall environmental quality.

National Estuary Program Established in 1987, this program's purpose is to identify, restore, and protect significant estuaries.

natural herbicides Any naturally occurring chemical that is destructive to weeds and other undesirable growth.

natural pesticides Any naturally occurring chemical that may be used to repel fungi, insects, and other pests.

natural selection The process, according to Darwin's theory of evolution, by which only the organisms of a species survive that best adapt to changes in the environment, each adaptation contributing to the evolution of that species.

nekton Free-swimming organisms.

NEPA (The National Environmental Policy Act of 1969) A U.S. law designed to protect and maintain overall environmental quality.

neutral Said of an aqueous solution with an even pH balance of 7.

neutron Atomic particle with no charge.

niche A species' place in the biotic world.

NIMBY (Not in my backyard) An attitude of tolerance toward a planned project as long as the individual holding the attitude is not directly and adversely affected.

nitrogen dioxide A highly reactive gas that comes from power plants and motor vehicles.

nitrogen oxides Substance formed when nitrogen reacts with oxygen.

Noise Control Act A 1972 act that serves to protect the health and well-being of U.S. citizens, wildlife and the environment by promoting a noise-free environment.

nonrenewable sources of energy Energy sources that exist in finite amounts.

nonsource pollution Pollution that tends to be comparatively more diffuse, less concentrated, and widespread.

Nonsource pollution Pollution that does not spring from one source.

nontransmittable disease A disease that cannot be spread through host contact.

NOAA The National Oceanic and Atmospheric Administration is the Federal agency that researches weather patterns, changes in our oceans, and in our climate.

no-till cultivation The farming practice that eliminates the use of tractors to be less invasive to the soil than till cultivation (using tractors).

not in my backyard (NIMBY) An attitude of tolerance toward a planned project as long as the individual holding the attitude is not directly and adversely affected.

nuclear fission The process of harvesting energy by splitting atoms.

nucleic acids Acids that make up DNA, the genetic blueprint of organisms.

observation Acquiring knowledge of the outside world by sensory perception.

oceanic trench Deep depressions and fissures produced by the collision and of two oceanic plates, one **subduct**ing under the other.

oceans Divisions of the Earth's World Ocean, these are vast expanses of highly salinated water (-marine habitats). They are generally considered the Pacific, the Atlantic, the Indian, and the Arctic.

old-growth forests Forests that have not been touched in 100 years or more.

oligotrophic Refers to aquatic environments with ample oxygen supply.

oligotrophic zones Lakes with ample supplies of oxygen.

OPEC nations Organization of Petroleum Exporting Countries.

ore A mineral with a high concentration of metal.

organ Systems formed by multiple tissue types with similar functions.

organ system Multiple organs working together for a specific function. The digestive system is an organ system made of the stomach, large intestines, and other organs.

organic molecules Molecules made up mostly of carbon and hydrogen.

organic molecules of life Fats, carbohydrates, proteins and DNA.

organism A whole sustained by multiple organ systems. An elephant, a mushroom, and an oak tree are examples of organisms.

OSHA (Occupational Health and Safety Act of 1970) A U.S. government administration created to "assure safe and healthful working conditions for working men and women by setting and enforcing standards and by providing training, outreach, education, and assistance."

overburden Waste that must be removed to retrieve coal in the **surface-mining** process.

overexploitation Excessive use of vital goods and resources taken from the earth faster than they are able to replenish naturally.

overkill Overhunting or overharvesting of species that endangers its population.

over-population A situation in which a particular population reaches numbers that its environment cannot maintain.

ozone (O_3) A compound found in high concentrations in the stratosphere (~10–30 miles from earth) that plays a critical role in protecting the earth from UV rays.

parasitism A type of symbiosis in which one species benefits while the other is harmed or sickened.

parasitoid Insect that uses a host to complete its life cycle, ultimately killing it.

particulates A pollutant composed of air-borne solids or liquids such as dust, smoke, or even brake pad debris that accumulates in the air we breathe.

partition of resources An activity within a community in which competing species will occupy a smaller niche to avoid directly competing for resources.

passive solar energy A means of harnessing sunlight without the use of a collector or solar cell.

pathogenicity The potential of a colony of bacteria or a virus to create disease.

pathogens Disease-causing microorganisms: usually bacteria, viruses, fungi, or protozoa.

pedosphere The entirety of the earth's sphere that encompasses soil.

permafrost A layer of continuously frozen ground littered with layers of decaying or dead leaves and **detritus**.

persistence Amount of time it takes for a substance to break down; tendency of compounds, such as ones containing fluorine, to remain in the atmosphere for extended periods.

petroleum A fossil fuel created from the fossils of living organisms that perished millions of years ago. This is a general term for the naturally occurring, flammable liquid found under geographic formations and its by-products.

pH A measure of how acidic or basic an aqueous solution measures. Its three basic categories are **acidic, neutral**, and **basic**.

photic zone The upper layer of the open seas. Whales are often found at this layer as they break the water surface to obtain oxygen.

photochemical smog An irritant to eyes, nose (called a lacrimal response), and to lungs that occurs when ozone reacts with sunlight.

photons Light acting as matter-like particles.

photovoltaics An active form of solar energy. Solar cells used to convert sunlight into electricity.

physical Addresses such factors in the environment as precipitation, temperature, and amount of sunlight.

phytoremediation A practice used to remove toxic chemicals from the soil through plants, since plants have the ability to absorb heavy metals through their roots.

pioneer growth The growth that occurs in the first 6–10 years in the cycle of change of an ecosystem, during which early growth occurs that facilitates growth for other plant species.

pioneer species The first organisms to grow after an initial disturbance.

pit mining A type of surface mining in which pits are dug to extract coal from seams found deep in a mountain.

planck response The ability to release infrared energy out of the upper atmosphere if it gets too warm to maintain global temperature stability.

plasmids Circular, noncoding DNA in bacteria that act as storehouses of DNA.

plate tectonics The slow movement of the outer layer of the earth.

pleiotropy The phenomenon in which a gene at one location codes for multiple expressions.

poaching The environmentally harmful activity of slaughtering animals for tusks, fur, or other valuable body parts. Poaching occurs when animal parts are worth a lot of money.

polar Describes the property of molecules having two distinct sides, each with distinct charges (positive or negative). In water molecules, the oxygen end of the molecule carries a net negative charge while the hydrogen end carries a net positive charge

pollution The result of chemicals and toxic waste from our labs and factories leaking into our soils, accumulating in the atmosphere, and leaching into our water supply.

pollution credits Credits from the government given to businesses that produce pollution less than their cap allows. These credits may be sold to companies that exceed their pollution cap or limit.

population A group of individuals of the same species living in proximity and interacting with one another.

potential energy Energy that is stored and ready to be released in some capacity.

poverty As used in ecological discussions, the widespread lack of means to be supplied with basic human needs. It commonly leads to overpopulation, disease, poor health care, and environmentally damaging practices such as poaching.

primary consumers Organisms that must consume organisms that exist at the primary trophic level for energy, retaining only 10% of the energy; Herbivore.

primary pollutants Pollution that is released directly into the atmosphere either by natural means such as volcanic activity and wildfire dust or by human activity such as automobile and coal factory pollution.

primary production Oil that bubbles to the surface after drilling as a result of the existing, naturally-occurring pressure from multiple layers of rock surrounding the reservoir.

primary production(2) Autotrophs assimilating carbon into their biomass.

primary sewage treatment Sewage treatment that uses a screen to filter out and extract water from solid waste. The solid waste that remains is called **sludge** and is eventually used as fertilizer.

primary succession vitalization that occurs in an ecosystem where there was previously, usually, bare rock and no soil or wildlife.

producers Organisms that produce their own energy and weigh in at the first tropic level. Synonym: **autotroph**.

profundal zone Zone of lake found near the bottom, no photosynthetic life found here.

prokaryotic cells Lack a nucleus and membrane-bound organelles.

proteins A diverse class of biological molecules with many functions which include comprising muscles, acting as enzymes, and enabling immune response.

proton Positively charged atomic particle.

provident ecology The assertion believed by the ancient Greeks that nature has an innate ability to maintain balance, so that any stimulus taking the earth away from its fulcrum would be counteracted with negative feedback to maintain stasis. By extension, if a species became overpopulated, thus threatening other species, nature would respond with an event such as disease that would decimate enough of the overpopulated species to protect itself and its biodiversity. Extinction of any species, therefore, could not occur.

quorum sensing A way of controlling bacterial pathogenicity that involves an analogy that blocks bacterial protein receptors.

radiation Type of cancer treatment involving the use of intense energy to target cancer cells.

rainforests Equatorial biomes that experience almost daily rainfall and little temperature fluctuation, and that display the earth's highest biodiversity.

range of tolerance The range in pH levels that aquatic life can survive.

realized niche A species' place in the biotic world; it is often much smaller than the fundamental niche. It is affected by **limiting factors** such as food availability, predators, competitors, and parasites.

rehabilitation Returning a damaged ecosystem to a state at which it is ecologically functional.

restoration The systematic, scientific approach carried out by humans to bring back a disturbed ecosystem to a healthy state.

restriction enzyme A type of "DNA scissor" that can find a specific, desired sequence of DNA and snip it from the "irrelevant" DNA.

retrovirus A virus that creates DNA from an RNA blueprint. HIV is an example of a retrovirus.

reverse osmosis The process in which water is pushed through a synthetic membrane whose pores are so miniscule that all impurities and contaminants are filtered out. Then it is exposed to UV light which destroys any remaining bacteria or viruses. Finally, it is treated with hydrogen peroxide.

rhizosphere A complex layer of the soil that comes in contact with plants where nutrient exchange occurs.

ribonucleic acid (RNA) Carrier of genetic codes that is made up of Nucleic Acids, made from DNA, and used as genetic information to make proteins.

riparian zones The outer banks of rivers and streams which contain rich, moist soil, and produce lush vegetation.

river The largest category of flowing water through a channel to or from lakes and oceans.

RNA (ribonucleic acid) Carrier of genetic codes that is made up of Nucleic Acids, made from DNA, and used as genetic information to make proteins.

rock A substance comprised of more than one mineral.

r-selected species Species that procreate in litters, do not put a lot of care into their young, have short life spans, and have evolved without significant competition for resources. Many rodents and insects are R-selected species.

rubisco Enzyme that a photosynthetic organism contains which has the ability to fix carbon, or convert it into a form that it can use to create carbohydrates. This allows herbivores that consume the organism to assimilate the carbon into its body.

runoff Precipitation that does not get fully absorbed by the ground and is lost as runoff as it moves downhill. With it, runoff can take the first few inches of nutrient-rich ground soil called **topsoil**. The flow of water strips away topsoil causing soil erosion, which can lead to **desertification**.

Safe Drinking Water Act A 1974 act passed is to ensure and protect the quality of U.S. drinking water mostly from underground sources.

sarcoma Type of cancer that originates in muscle and connective tissue.

second law of thermodynamics When energy changes form, most of the energy is lost as heat.

secondary consumers Carnivores that consume herbivores for energy.

secondary pollutants Substances such as ozone, nitric acid, and sulfuric acid caused by the chemical changes formed when **primary pollutants** react with other pollutants in the atmosphere.

secondary production Oil that bubbles to the surface after **primary production** as a result of further drilling or pressure produced by gas pumped into the reservoir within the rock.

secondary production(2) Heterotrophs assimilating carbon into their biomass.

secondary sewage treatment Sewage treatment stage during which bacteria is added to the filtered water in order to break down any remaining organic matter.

secondary succession Revitalization that occurs where a previous ecosystem was disturbed by such causes as fire, deforestation, over-grazing, etc.

sedimentary rock Rock formed when sediment accumulates and mineralizes. Examples of sedimentary rock include limestone and sandstone.

seed banks Storage facilities where seeds are stored in low temperatures and low humidity. Also known as **gene banks.**

selective cutting Forest cutting in which only the most desirable trees are isolated and cut down.

sensitive species A species that is vulnerable to habitat loss.

sludge The solid waste that remains from **primary sewage treatment** that is eventually used as fertilizer.

solifluction The process of solid water-saturated soil moving down a frozen slope to a lower altitude, taking debris with it.

source pollution Pollution emitted directly into a surface through a confined, direct means such as a pipe, drain, or sewage from one source.

specialist species Species that occupy a narrow niche. They are sensitive to environmental change.

speciation The process of a new species forming as a result of adaptation to an ever-changing environment.

species abundance The number of individuals of a given species represented in a given area.

species diversity The number of different species (see species richness) as well as the differences among and within a given ecological community.

species richness The total number of different species present in a given area.

sticky ends The result of isolating a relevant sequence of a DNA trait from a particular species using restrictive enzymes to make it complimentary to a desired plasmid DNA.

stratosphere The second atmospheric layer which extends from 10 to 30 miles outward. It contains high concentrations of ozone (O_3), which prevents UV light from entering earth's lower atmosphere.

streams Any flow of water through a channel smaller than a river.

strip cutting Tree cutting in which a corridor is cut through a patch of forest land.

strip farming Same as **alley farming.** The practice of growing crops in rows to keep pests confused and to maintain soil fertility. Rows of crops are also typically flanked by tall vegetation in order to block excess wind, thereby preventing erosion.

subduct Literally, move under. In **plate tectonics**, usually an older and thinner plate slides under the other more substantial plate.

Subduction The collision of an oceanic and a continental plate that creates a **convergent boundary.**

subsistence farming Farming for the purpose of feeding one's family. Farming to live. subsistence farming Farming for the purpose of feeding one's family. Farming to live.

subsurface mining Mining that utilizes shafts to retrieve minerals deep under the earth's surface.

succession The sequence of events that occur after an ecosystem disturbance.

sulfur dioxide A colorless gas with a noxious odor released as a result of fossil fuel combustion.

sulfuric acid (H_2SO_4) Acid sometimes formed when sulfur dioxide reacts with certain molecules that are present. Sulfuric acid is a corrosive compound that is in part responsible for acid rain.

surface fires Cool forest fires that provide many ecological benefits such as the release of nutrients from the soil, germination of plants, and control of pest populations.

surface mining Coal mining that removes waste (**overburden**) to reach coal that is to be retrieved.

sustainability Practice of living in such a way that vital goods are taken from the earth conservatively, avoiding long-term damage or depletion.

symbiosis Different species of organisms living together over an extended period.

sympatric speciation Speciation which occurs without a physical barrier.

taiga (boreal forests) These biomes are vast spruce, fir, and pine forests in colder climates in the northern hemisphere.

telomerase Enzyme that restores the ends of DNA.

telomeres Noncoding ends of DNA.

temperature inversion A reversal of normal temperature patterns resulting from warm, light air moving over cold, dense air, the cool air being trapped below a column of warm air. Pollution gets trapped below the inversion layer and cannot dissipate easily into the atmosphere.

teratogen agent That causes birth defects by effecting fetal development.

terrace farming Farming on steep slopes, with steps cut into the slopes in order to maintain water and to keep the ground covered with vegetation (in order to prevent runoff).

terrestrial ecosystem An ecosystem that exists on land, rather than in water.

tertiary consumers Carnivores that eat other carnivores. Lions and hyenas are tertiary consumers.

tertiary sewage treatment The final stage of sewage treatment during which UV light is used to kill off any remaining bacteria and viruses after which treatment with chlorine makes it become potable.

TFR (total fertility rate) the number of children born to a woman of reproductive age.

thermal energy Measured in Joules, thermal energy is produced by either an increase or decrease in temperature.

thermal stratification Temperature change with depths of a body of water.

thermosphere The largest part of the earth's atmosphere extending from 50 miles to 300 miles outward from earth's surface.

threatened species A species that has a good chance of becoming endangered in the near future.

tissue A group of different cell types that support each other for a common function.

topsoil The first few inches of nutrient-rich ground soil: soil composed of air, water, minerals, and an abundance of organic microorganisms.

total fertility rate (TFR) The number of children born to a woman of reproductive age.

toxic chemicals Chemicals that may cause either long or short term harm to humans and animals exposed to such chemicals.

trace element Elements such as iron and magnesium that are not part of **CHNOPS** but are needed by the body in small doses.

transform fault Separation of plates caused by the enormous pressure of the two grinding together, causing high earthquake activity.

transgenic Having DNA that derives from two or more sources.

transgenic crops Crops with DNA from two or more sources.

transpiration Process that occurs when water exits pores in the leaves of plants and trees and then enters the atmosphere as water vapor.

trophic levels Positions of organisms in the food chain.

tropical rainforests Equatorial biomes with almost daily rainfall and characterized by the earth's highest biodiversity and little temperature fluctuation.

troposphere The atmospheric layer closest to earth that extends from earth's surface to roughly ten miles outward. The troposphere is made up of gases such as oxygen, carbon dioxide, and nitrogen.

tundra A terrestrial biome characterized by permafrost, long winters, and short growing seasons.

umbrella species A species that, when conserved, also conserves species that share its ecosystem. An example is the Northern Spotted Owl.

UNEP (United Nations Environment Programme) The United Nations' voice regarding the environment.

United Nations Environment Programme (UNEP) The United Nations' voice regarding the environment.

UNFCC United Nations Framework Convention on Climate Change, the convention committed to reducing carbon emissions and mitigating effects of climate change.

variable A factor in a hypothesis that is to be adjusted or manipulated.

vector An agent, such as a plasmid, used to carry a DNA trait to another cell, such as bacteria in the case of plasmids.

vestigial structure A body part no longer used by the organism but left over from a time when it was.

vigorous flowering The activity that occurs, approximately 7–15 years into the cycle of change of an ecosystem.

VOC's An acronym for **volatile organic compounds** which are organic molecules that transform into a gas at room temperature.

volatile organic compounds (VOC's) Organic molecules that transform into a gas at room temperature.

vulnerable species Faces imminent extinction unless conservation action is taken.

water table A surface layer of water at which the pressure of groundwater equals atmospheric pressure. The water table is often observed as a flat or level surface of water.

watershed An aquatic network where local surface water drains and eventually effuses into a single point or waterway.

wavelengths The distance between the crests of waves of electromagnetic energy.

waves The means by which electromagnetic energy travels through space.

weather Variation in precipitation, cloud cover, and temperature on the earth's surface that results from unequal solar warming.

wildlife corridors Connections of wildlife habitat patches created by conservationists.